Lecture Notes in Mathematics　　2042

Editors:
J.-M. Morel, Cachan
B. Teissier, Paris

W0234393

For further volumes:
http://www.springer.com/series/304

William J. Layton • Leo G. Rebholz

Approximate Deconvolution Models of Turbulence

Analysis, Phenomenology and Numerical Analysis

 Springer

William J. Layton
University of Pittsburgh
Dept. Mathematics
Pittsburgh Pennsylvania
USA

Leo G. Rebholz
Clemson University
Department of Mathematical Sciences
Clemson South Carolina
USA

ISBN 978-3-642-24408-7   e-ISBN 978-3-642-24409-4
DOI 10.1007/978-3-642-24409-4
Springer Heidelberg Dordrecht London New York

Lecture Notes in Mathematics ISSN print edition: 0075-8434
                              ISSN electronic edition: 1617-9692

Library of Congress Control Number: 2011943497

Mathematics Subject Classification (2010): 65-XX, 76-XX

Printed on acid-free paper

Springer is part of Springer Science+Business Media (www.springer.com)

Contents

Chapter 1
Introduction

This book presents a mathematical development of a recent approach to the modeling and simulation of turbulent flows based on methods for the approximate solution of inverse problems. The resulting Approximate Deconvolution Models or ADMs have some advantages (as well as some disadvantages) over more commonly used turbulence models:

- ADMs are supported by a mathematically rigorous theoretical foundation.
- ADMs are a family of models of increasing accuracy $O(\delta^{2N+2})$, where δ is the averaging (or filter) radius.
- To the extent that the phenomenology of turbulence is understood, ADMs have been shown to give correct predictions of turbulent flow statistics.
- The ADM microscale can, by a judicious selection of the model's one parameter, be made to equal the model's filter radius.
- The whole family of models conserves (in the appropriate context) all the integral invariants of the Euler equations.
- With smooth data, ADMs have unique, smooth, strong solutions which converge to a weak solution of the NSE (modulo a subsequence) as the filter radius decreases and which approach a smooth global attractor as time increases.
- The abstract theory of the model uncouples "stability" of the models from "accuracy" of the models. This abstract theory thus gives a path for the development of still better models by isolating the properties needed in deconvolution operators for increased accuracy (smaller consistency error for turbulent velocities) within the operators that give a robust theory.

The approach of ADMs does not solve all the problems of turbulent flow simulation. In particular, we note that there are many problems remaining in the theory and practice of ADMs:

- The problem of commutator errors which arises due to either filtering through a boundary or changing the averaging radius remains. It may

W.J. Layton and L.G. Rebholz, *Approximate Deconvolution Models of Turbulence*, Lecture Notes in Mathematics 2042, DOI 10.1007/978-3-642-24409-4_1, © Springer-Verlag Berlin Heidelberg 2012

even be exacerbated because the standard approach to deconvolution is
based on repeated filtering.

- Put another way, ADMs approximate in a direct way the space filtered
  NSE. Thus, the space filter of the divergence of the stress tensor must be
  calculated somehow. While boundary conditions for velocities at walls are
  known, no similar knowledge of stress at walls can be used to simplify this
  calculation.
- The problem of boundary conditions for flow averages remains as difficult
  as ever. At this point the most promising approach seems to be variable
  averaging radius which approaches zero at walls. The results we present
  herein for constant δ are largely open for variable δ.
- The mathematical foundation for van Cittert deconvolution implicitly uses
  that the transfer function of the filter satisfies $0 \leq \widehat{g}_\delta(k) \leq 1$. However,
  many filters chosen due to practical issues violate this. Thus, it seems that
  new deconvolution methods are needed that work well for more general
  filters.
- Even in the simplest case of differential filters, a clean numerical discretization of a full ADM involves solving a fourth order problem, which increases
  complexity and adds to the issue of numerical boundary conditions. The
  (small) extra terms can be treated by numerical artifices, increasing
  efficiency as well as numerical noise.
- Linear time relaxation, while a wonderfully elegant mathematical and
  algorithmic idea, may not suffice to damp sufficiently noise coming from
  numerical errors, the above issues and other effects. One can hope that as
  the above issues are resolved and as numerics improve, this may lessen.
  However, there is a parallel in wave number space between linear time
  relaxation and linear artificial viscosity. Thus, more effective and more
  selective (thus nonlinear) realizations of time relaxation are needed.

A comprehensive analytical theory of approximate deconvolution models
(ADMs) in large eddy simulation (LES) begins now to take shape for the
most favorable or best understood combination of *van Cittert deconvolution,
differential filters, constant averaging radius,* and *away from walls* (meaning
for periodic boundary conditions). Open theoretical questions (strongly
connected to computational difficulties) abound for variable $\delta = \delta(x)$,
near wall behavior and models and filters used in practice. Possibly these
difficulties are linked to the use of the van Cittert deconvolution procedure.
(They have provided motivation for development of both deconvolution
regularizations of the NSE and other deconvolution methods.) If serious
issues arise in filtering near walls then repeated filtering (as performed by
van Cittert) should not be expected to improve them.

A clear outline of a theory, even in the simple case herein, can be very
useful in approaching the hard problems that remain. We have developed
the basic results of this theory herein. Our goal is to provide a clear
and complete mathematical development of ADMs (in the case where a

clear and complete development is possible) to ease entry of applied and computational mathematicians into the area for its further development, addressing the above noted difficulties. We present herein the analytical theory of ADMs with its connections, motivations and complements in the phenomenology of and algorithms for ADMs.

1.1   The Navier–Stokes Equations

"As far as I can see, there is today no reason not to regard the hydrodynamic equations as the exact expression of the laws that rule the motion of real fluids."
   H. Helmholtz, 1873, p. 158 in: *Über ein Theorem, geometrische ähnliche Bewegungen flüssigkeiten*, Königliche Akademie der Wissenschaffen zu Berlin, Monatsberichte, p. 501–514.

The motion of an incompressible, viscous fluid is completely described by the Navier–Stokes equations (NSE). In a region Ω in $\mathbb{R}^2$ or $\mathbb{R}^3$, driven by a body force $f(x,t)$, the fluid velocity and pressure satisfy:

$$u_t + u \cdot \nabla u - \nu \Delta u + \nabla p = f, \ x \in \Omega, \ 0 < t \leq T, \tag{1.1}$$

$$\nabla \cdot u = 0, \ x \in \Omega, \ 0 < t \leq T, \tag{1.2}$$

$$u(x,0) = u_0(x), \ x \in \Omega, \ \text{and} \tag{1.3}$$

$$\int_\Omega p \ dx = 0, \ 0 < t \leq T. \tag{1.4}$$

The zero-mean constraint on the pressure is a commonly used normalization condition that fixes the undetermined additive constant. The typical boundary conditions studied are either (a) the *no slip condition* which holds on fixed, solid walls,

$$u = 0 \text{ on } \partial\Omega,$$

or L periodic *(with zero mean)* boundary conditions:

$$u(x + Le_j, t) = u(x,t) \quad j = 1,2,3, \tag{1.5}$$

$$\int_\Omega \phi \ dx = 0 \quad \text{for} \quad \phi = u, \, u_0, \, f, \, p, \, P.$$

The L periodic boundary condition has the virtue of separating the difficulties of the NSE from possibly harder difficulties of the interaction of flows with walls. Throughout, we shall denote the usual $L^2(\Omega)$ norm (which measures the kinetic energy in a velocity) and inner product by

$$||u|| := \sqrt{\int_\Omega |u(x)|^2 dx}, \text{ and } (u,v) = \int_\Omega u \cdot v dx.$$

Although the NSE are derived directly from conservation laws [L08] and have been known for over 150 years, their mathematical understanding remains incomplete. Even in the simplest cases, such as periodic domains with infinitely smooth data, determining whether the NSE are well-posed is a Clay prize problem [Fe00].

The complexity of solutions, richness of their scales, difficulty in their numerical simulation and sensitivity to data, discretization and scale truncation errors is roughly measured by the Reynolds number, e.g., [R90],

$$Re := \frac{VelocityScale \cdot LengthScale}{KinematicViscosity} = \frac{UL}{\nu}.$$

In the definition of Re the length scale L is chosen to represent the size of the largest flow structures. When applying dynamic similarity it only has meaning when used to relate two geometrically similar domains with L chosen the same way for both. At this level of generality we will most often take L to be a characteristic length of the domain, as in (1.5) for periodic boundary conditions or $diam(\Omega)$ for non-periodic ones. For specific problems these are often not the most insightful choices. For large Reynolds numbers in $3d$, persistent, energetically significant scales exist down to the Kolmogorov microscale η of

$$\eta = \text{Re}^{-3/4} L.$$

The key to making progress in the mathematical understanding of the Navier–Stokes equations is the energy equality. If u, p is a smooth solution to (1.1)–(1.4), then multiplying the momentum equation by u, integrating over Ω, dividing by the volume of Ω, denoted $|\Omega|$, to scale out the volume dependence, integrating by parts and integrating in time gives

$$\frac{1}{2|\Omega|}||u(t)||^2 + \int_0^t \frac{\nu}{|\Omega|}||\nabla u(t')||^2 dt' = \frac{1}{2|\Omega|}||u_0||^2 + \int_0^t \frac{1}{|\Omega|}(f(t'), u(t'))dt'.$$

The energy equality describes the evolution of the kinetic energy (per unit volume and mass). There are three terms involved:

$$\text{The kinetic energy}: \ E(t) := \frac{1}{2|\Omega|} \int_\Omega |u(x,t)|^2 \, dx,$$

$$\text{The energy dissipation rate}: \ \varepsilon(t) := \frac{\nu}{|\Omega|} \int_\Omega |\nabla u(x,t)|^2 \, dx,$$

$$\text{Energy input via body force flow interaction} := \frac{1}{|\Omega|} \int_\Omega f \cdot u \, dx,$$

where ν is the kinematic viscosity. For strong solutions of the NSE the above energy equality holds. For weak solutions, the best that is known is the energy inequality

$$E(t) + \int_0^t \varepsilon(t) \, dt \le k(0) + \int_0^t \frac{1}{|\Omega|} \int_\Omega f \cdot u \, dxdt.$$

1.1.1   Integral Invariants

"The problem of course is that there are certain physical integral constraints which are deducible from the continuous hydrodynamic equations such as the conservation of total westerly angular momentum for smooth boundaries. How does one know when one is conserving total angular momentum in finite differences? How does one measure it?

This brings us to a related question. Suppose there are a number of conservation principles deducible, e.g. that of total angular momentum and of total energy. However these principles apply to different powers of the velocity. Is it possible to devise a differencing scheme and a corresponding quadrature scheme which will simultaneously preserve both of the integral constraints irrespective of which continuous system is integrated numerically?" Joseph Smagorinsky in: Frontiers of Numerical Mathematics, (R.E. Langer, editor), University of Wisconsin Press, Madison, 1060.

"One can also verify that the total kinetic energy of the liquid remains bounded but it does not seem possible to deduce from this fact that the motion itself remains regular." J. Leray, in: [L34a].

"The *uniqueness* of the flow and *the continuity* of its kinetic energy at the moments of time where it is not regular *are two open problems*; they are related." J. Leray, 1954, in [L54].

Using a vector identity the NSE can be rewritten with its nonlinearity in *rotational form*, which makes its integral invariants clearer

$$u_t + (\nabla \times u) \times u - \nu \triangle u + \nabla P = f(x,t), \qquad (1.6)$$

$$\text{and} \ \ \nabla \cdot u = 0,$$

$$\text{where} \ \ P = p + \frac{1}{2}|u|^2. \qquad (1.7)$$

Definition 1. The Energy, Helicity and Enstrophy of a fluid (per unit volume) are given by

$$E(t) := \frac{1}{|\Omega|} \int_\Omega \frac{1}{2}|u(x,t)|^2 dx,$$

$$H(t) := \frac{1}{|\Omega|} \int_\Omega \frac{1}{2}(\nabla \times u(x,t)) \cdot u(x,t) dx,$$

$$Enstrophy(t) := \frac{1}{|\Omega|} \int_\Omega |\nabla \times u(x,t)|^2 dx.$$

Helicity is a fundamentally important quantity in turbulent fluid flow that provides important information on the topology of vortex filaments [MT92]. Helicity density has been used as an indicator of coherent structures, and moreover, as an indicator of low turbulent dissipation. The Pythagorean identity gives

$$\frac{(u \cdot \nabla \times u)^2 + |u \times \nabla \times u|^2}{|u|^2 |\nabla \times u|^2} = 1, \tag{1.8}$$

which can be interpreted as

$$\frac{Helicity^2 + NSEnonlinearity^2}{Energy \cdot Enstrophy} = 1. \tag{1.9}$$

Hence at constant energy and enstrophy, high helicity implies small nonlinearity, and conversely. The interaction between model's treatment of energy and helicity impacts its predictions of a flow's rotational structures.

Theorem 2. *Let u be a sufficiently smooth solution of the homogeneous Euler equations (the NSE with $\nu = 0$) and under L periodic boundary conditions. Then, for all t, the energy in both 2d and 3d flows and helicity in 3d flows satisfy*

$$E(t) = E(0),$$

$$H(t) = H(0).$$

For 2d flows the enstrophy satisfies

$$Enstrophy(t) = Enstrophy(0).$$

Proof. Let the vorticity be denoted as usual by $\omega = \nabla \times u$. Note that

$$(\omega \times u) \cdot u = 0 \text{ and } (\omega \times u) \cdot \omega = 0.$$

For conservation of energy, take the dot product of the NSE in rotational form with u. This gives

$$\frac{d}{dt}E(t) = 0.$$

For helicity, note that the curl operator is self-adjoint:

$$\int_\Omega (\nabla \times u) \cdot v \, dx = \int_\Omega u \cdot (\nabla \times v) dx.$$

Thus $(u_t, \omega) = \frac{1}{2}\frac{d}{dt}H(t)$. Now take the inner product of the NSE in rotational form with ω and use the last two observations. For conservation of enstrophy in 2d, recall that the 2d vorticity is a scalar and the 2d vorticity transport equation is

$$\omega_t + u \cdot \nabla\omega - \nu\triangle\omega = \nabla \times f.$$

With $\nu = f = 0$ take the inner product with ω. This proves conservation of enstrophy in 2d. □

When $\nu > 0$ there are two additional important integral quantities for the Navier–Stokes equations in 3d and one in 2d

$$\text{Energy dissipation rate } \varepsilon(t) = \frac{\nu}{|\Omega|} \int_\Omega |\nabla u(x,t)|^2 dx,$$

$$\text{Helicity dissipation rate } \gamma(t) = \frac{\nu}{|\Omega|} \int_\Omega (\nabla \times \omega(x,t)) \cdot \omega(x,t) dx,$$

$$\text{Enstrophy dissipation rate(2d)} = \frac{\nu}{|\Omega|} \int_\Omega |\nabla \times \omega(x,t)|^2 dx.$$

The energy dissipation rate can be defined equivalently by $\varepsilon(t) = \frac{2\nu}{|\Omega|} \int_\Omega D_{ij} D_{ij} dx$ where $D_{ij} = (1/2)(u_{i,j} + u_{j,i})$ is the deformation tensor (symmetric part of the velocity gradient tensor) and the summation convention is used.

1.1.2 The K41 Theory of Homogeneous, Isotropic Turbulence

"The observational material is so large that it allows to foresee rather subtle mathematical results, which would be very interesting to prove." A.N. Kolmogorov, 1978, in: *Remarks on statistical solutions of the Navier–Stokes equations*, Uspekhi Mate. Nauk 33, p. 124.

"When I die and go to Heaven there are two matters on which I hope enlightenment. One is quantum electrodynamics and the other is turbulence. About the former, I am really rather optimistic." H. Lamb, 1932, quoted in: Computational Fluid Mechanics and Heat Transfer, by Anderson, Tannehill and Pletcher, 1984.

Turbulent flow is observed to consist of a cascade of three dimensional eddies of various sizes. Why do solutions of the Navier–Stokes equations exhibit this energy cascade? The answer to this question has been understood since the work of L. F. Richardson and A. N. Kolmogorov and is based on a few fundamental properties of solutions of the Navier–Stokes equations:

- If $\nu = 0$ the total kinetic energy of the flow is exactly equal to the total kinetic energy input by body forces:[1]

$$E(u)(t) = E(u)(0) + \int_0^t \frac{1}{L^3} \int_\Omega f \cdot u \, dx \, dt.$$

- The nonlinearity conserves energy globally (since $\int_\Omega u \cdot \nabla u \cdot u dx = 0$) but acts to transfer energy to smaller scales by breaking down eddies into smaller eddies (for example, if $u \simeq (U \sin(\frac{\pi x_1}{l}), 0, 0)^{tr}$ has wave length l and frequency $\frac{\pi}{l}$ then $u \cdot \nabla u \simeq \frac{U^2 \pi}{2l}(\sin(\frac{\pi x_1}{l/2}), 0, 0)^{tr}$ has shorter wave length $\frac{l}{2}$).

- If $\nu > 0$, since $\varepsilon(u)(t') \geq 0$ the viscous terms dissipate energy from the flow globally:

$$E(u)(t) + \int_0^t \varepsilon(u)(t')dt' = E(u)(0) + \int_0^t \frac{1}{L^3} \int_\Omega f \cdot u \, dx \, dt.$$

- For Re large the energy dissipation due to the viscous terms is negligible except on very small scales of motion. For example, if $u \simeq (U \sin(\frac{\pi x_1}{l}), 0, 0)^{tr}$ then *on this scale*,

$$\text{viscous term} = -\nu \triangle u \simeq \pi^2 \frac{\nu U}{l^2} < \sin\left(\frac{\pi x_1}{l}\right), 0, 0 >^T$$

$$\text{energy dissipation} = \varepsilon(u) \simeq \frac{C}{L^3} \frac{\nu U^2}{l^2}.$$

---

[1]For the physical reasoning in this section it is appropriate to suppose that the energy equality holds and sidestep the deeper questions concerning weak vs. strong solutions and energy equality vs. energy inequality, e.g., [Ga00, Gal94].

Thus the nonlinear term dominates and the viscous term is negligible if

$$\frac{U^2}{l} >> \frac{\nu U}{l^2} \text{ , i.e., } \frac{lU}{\nu} >> 1.$$

- The forces driving the flow input energy persistently into the largest scales of motion.

The picture of the energy cascade that results from these effects is thus:

- Energy is input into the largest scales of the flow.
- There is an intermediate range in which nonlinearity drives this energy into smaller and smaller scales and conserves the global energy because dissipation is negligible.
- Eventually, at small enough scales, dissipation is non negligible and the energy in those smallest scales decays to zero exponentially fast.

In 1941 A. N. Kolmogorov gave a quantitative description of the energy cascade (often called the K41 theory). The most important components of the K41 theory are the time (or ensemble) averaged energy dissipation rate, ε, and the distribution of the flows averaged kinetic energy across wave numbers, $E(k)$. Let $< \cdot >$ denote long time averaging so that the (time averaged) energy dissipation rate of that flow is thus defined to be

$$< \varepsilon > := \lim_{T \to \infty} \sup \int_0^T \frac{1}{L^3} \int_\Omega \nu |\nabla u(x,t)|^2 dx \, dt.$$

To present the K41 theory, some preliminaries are necessary.

1.1.2.1   Fourier Series

Consider the NSE under L periodic boundary conditions including the zero mean condition

$$\int_\Omega \phi \, dx = 0 \text{ on } \phi = u, p, f, \text{ and } u_0.$$

We can thus expand the fluid velocity in a Fourier series

$$u(x,t) = \sum_{\mathbf{k}} \widehat{u}(\mathbf{k}, t) e^{-i\mathbf{k} \cdot x}, \text{ where } \mathbf{k} = \frac{2\pi \mathbf{n}}{L} \text{ is the wave number and } \mathbf{n} \in \mathbb{Z}^3.$$

The Fourier coefficients are given by

$$\widehat{u}(\mathbf{k}, t) = \frac{1}{L^3} \int_\Omega u(x,t) e^{-i\mathbf{k} \cdot x} dx.$$

Magnitudes of $\mathbf{k}, \mathbf{n}$ are defined by

$$|\mathbf{n}| = \{|n_1|^2 + |n_2|^2 + |n_3|\}^{\frac{1}{2}}, k := |\mathbf{k}| = \frac{2\pi|\mathbf{n}|}{L},$$

$$|\mathbf{n}|_\infty = \max\{|n_1|, |n_2|, |n_3|\}, |\mathbf{k}|_\infty = \frac{2\pi|\mathbf{n}|_\infty}{L}.$$

The length-scale of the wave number $\mathbf{k}$ is defined by $l = \frac{2\pi}{|\mathbf{k}|_\infty}$. The energy in the flow can be decomposed by wave number as follows. For $u \in L^2(\Omega)$,

$$\frac{1}{L^3} \int_\Omega \frac{1}{2}|u(x,t)|^2 dx = \sum_\mathbf{k} \frac{1}{2}|\widehat{u}(\mathbf{k},t)|^2 = \sum_k \left(\sum_{|\mathbf{k}|_\infty = k} \frac{1}{2}|\widehat{u}(\mathbf{k},t)|^2 \right),$$

$$\text{where } \mathbf{k} = \frac{2\pi\mathbf{n}}{L} \text{ is the wave number and } \mathbf{n} \in \mathbb{Z}^3.$$

Definition 3. The kinetic energy distribution functions are defined by

$$E(k,t) = \frac{L}{2\pi} \sum_{|\mathbf{k}|_\infty = k} \frac{1}{2}|\widehat{u}(\mathbf{k},t)|^2, \text{ and}$$

$$E(k) := < E(k,t) >,$$

Parseval's equality thus can be rewritten as

$$\frac{1}{L^3} \int_\Omega \frac{1}{2}|u(x,t)|^2 dx = \frac{2\pi}{L} \sum_k E(k,t), \text{ and}$$

$$< \frac{1}{L^3} \int_\Omega \frac{1}{2}|u(x,t)|^2 dx >= \frac{2\pi}{L} \sum_k E(k).$$

1.1.2.2   The Inertial Range

"For the locally isotropic turbulence the [velocity fluctuation] distributions F_n are uniquely determined by the quantities ν, the kinematic viscosity, and ε, the rate of average dispersion of energy per unit mass [energy flux]..... For pulsations [velocity fluctuations] of intermediate orders where the length scale is large compared to the scale of the finest pulsations, whose energy is directly dispersed into heat due to viscosity, the distribution laws F_N are uniquely determined by F and do not depend on ν." I. Kolmogorov, 1941, (Statement of the first and second hypothesis of similarity), in: *The local structure of turbulence in incompressible viscous fluids for very large Reynolds number*, Dokl. Akad. Nauk SSSR 30 (1941), 9–13.

The K41 theory states that at high enough Reynolds numbers there is a range of wave numbers

$$0 < k_{\min} := U\nu^{-1} \leq k \leq\, < \varepsilon >^{\frac{1}{4}} \nu^{-\frac{3}{4}} =: k_{\max} < \infty, \qquad (1.10)$$

known as the inertial range, beyond which the kinetic energy in a turbulent flow is negligible, and in this range

$$E(k) \doteq \alpha < \varepsilon >^{\frac{2}{3}} k^{-\frac{5}{3}}, \qquad (1.11)$$

where α is the universal Kolmogorov constant whose value is generally believed to be between 1.4 and 1.7, k is the wave number and ε is the particular flow's energy dissipation rate. In this formula, the energy dissipation rate $< \varepsilon >$ is the only parameter which differs from one flow to another. We recall the argument leading to this description of the inertial range.

Conjecture 4. There is a range of wave-numbers, called the inertial range, over which (up to negligibly small effects) the time averaged kinetic energy only depends on the time averaged energy dissipation rate ε and the wave number k.

Beginning with this, postulate a simple power law dependency of the form

$$E(k) \simeq C < \varepsilon >^{a} k^{b}. \qquad (1.12)$$

If this relation is to hold the units, denoted by $[\cdot]$ on the LHS must be the same as the units on the RHS, $[LHS] = [RHS]$. The three quantities in the above have the units[2]

$$[k] = \frac{1}{length}, [<\varepsilon>] = \frac{length^2}{time^3}, [E(k)] = \frac{length^3}{time^2}.$$

Inserting these units into the above relation gives

$$\frac{length^3}{time^2} = \frac{length^{2a}}{time^{3a}} \frac{1}{length^b} = length^{2a-b} time^{-3a}, \text{ giving}$$

$$3a = 2, 2a - b = 3, \text{ or } a = \frac{2}{3}, b = -\frac{5}{3}.$$

---

[2]Note that from the definitions of E and $E(k)$ we have $[E] = \frac{length^2}{time^2}$, and $[E(k)] = length \times [E] = \frac{length^3}{time^2}$.

Thus, Kolmogorov's law follows

$$E(k) = \alpha < \varepsilon >^{\frac{2}{3}} k^{-\frac{5}{3}},$$

over the inertial range within $0 < k \leq C(LRe^{-\frac{3}{4}})^{-1}$.

The above estimate $\eta \sim LRe^{-\frac{3}{4}}$ for the Kolmogorov micro-scale is derived by similar physical reasoning. Let the reference large scale velocity and length (which are used in the definition of the Reynolds number) be denoted by U, L. At the scales of the smallest persistent eddies (the bottom of the inertial range) we shall denote the smallest scales of velocity and length by v_{small}, η. We form two Reynolds numbers:

$$Re = \frac{UL}{\nu}, Re_{small} = \frac{v_{small}\eta}{\nu}.$$

The global Reynolds number measures the relative size of viscosity on the large scales and when Re is large the effects of viscosity on the large scales are then negligible. The smallest scales Reynolds number similarly measures the relative size of viscosity on the smallest persistent scales. Since it is non-negligible we must have

$$Re_{small} \simeq 1, \text{ equivalently } \frac{v_{small}\eta}{\nu} \simeq 1.$$

Next comes an assumption of statistical equilibrium:

Energy Input at large scales = Energy dissipation at smallest scales.

The largest eddies have energy which scales like $O(U^2)$ and associated time scale $\tau = O(\frac{L}{U})$. The rate of energy transfer/energy input is thus $O(\frac{U^2}{\tau}) = O(\frac{U^3}{L})$.[3] The small scales energy dissipation from the viscous terms scales like

$$\varepsilon_{small} \simeq \nu |\nabla u_{small}|^2 \simeq \nu \left(\frac{v_{small}}{\eta}\right)^2.$$

Thus we have the second ingredient:

$$\frac{U^3}{L} \simeq \nu \left(\frac{v_{small}}{\eta}\right)^2.$$

Solving the first equation for v_{small} gives $v_{small} \simeq \frac{\nu}{\eta}$. Inserting this value for the small scales velocity into the second equation, solving for the length-scale

---

[3]It is known for many turbulent flows that, as predicted by K41, ε scales like $\frac{U^3}{L}$. This estimate expresses statistical equilibrium in K41 formalism, [F95, Po00].

η and rearranging the result in terms of the global Reynolds number gives the following estimate for η which determines the above estimate for the highest wave-number in the inertial range:

$$\eta = \eta_{Kolmogorov} \simeq Re^{-\frac{3}{4}} L.$$

This estimate for the size of the smallest persistent solution scales is the basis for the estimates of $O(Re^{\frac{9}{4}})$ mesh-points in space leading to complexity estimates of $O(Re^3)$ for DNS of turbulent flows.

1.2   Large Eddy Simulation

When (as usually occurs) it is not computationally feasible within time and resource limitations to solve the NSE on a computational mesh resolving eddies down to the microscale, the obvious solution, which is the aim of Large Eddy Simulation (LES), is to solve for flow averages instead. To begin, suppose we have some estimate of a feasible meshwidth $\triangle x$. From this a filter radius, traditionally denoted δ because of its association with the meshwidth, is selected, such as $\delta = \triangle x$, $\delta = 6\triangle x$ or $\delta = \eta_{\text{model}}$. Then a filter is selected with this filter radius (denoted by overbar). Most filters used are linear averaging operators that commute with differentiation in the absence of boundaries. With an averaging operator defined, the fluid velocity can be decomposed into means $\overline{u}$, which are computable/observable on a given mesh, and fluctuations $u' := u - \overline{u}$, which are unknowable on a given mesh or numerical simulation, by

$$u = \overline{u} + u'.$$

The filtered part $\overline{u}$ is also called the resolved velocity. With such a filter, averaging the NSE gives the exact Space Filtered NSE or SFNSE

$$\overline{u}_t + \overline{u \cdot \nabla u} - \nu \triangle \overline{u} + \nabla \overline{p} = \overline{f}(x,t), \text{ and } \nabla \cdot \overline{u} = 0.$$

The large structures $\overline{u}$ (defined by filtering) are time dependent and contain most of the flow's kinetic energy and are the targets of LES. There are several distinctions between LES and RANS. These large structures are time dependent so LES solves evolution equations for $\overline{u}(x,t)$. In LES the filter radius δ is related to computational resources and is decreasing as computational resources increase. Thus the modeling question is to model the effects of u', which is small in that it contains little kinetic energy, upon $\overline{u}$, which is large in the same sense. Thus in LES simple models (of the effects of a small variable upon a large one) are possible which in RANS highly accurate models of the effects of some O(1) quantities upon other O(1) quantities are needed. Finally, if δ is in the inertial range, then away from walls, time averages of u' are expected to have a universal structure which can be exploited to develop accurate and more universal models.

One fundamental issue in LES is closure. The closure problem is to write the non closed term in the SFNSE

$$\overline{u \cdot \nabla u} = \nabla \cdot (\overline{u\,u})$$

in terms of $\overline{u}$ not u. The closure problem is typically ill posed and thus fundamentally impossible to solve exactly in a useful way. The issues are then how to solve it approximately and what is the test for a useful solution? Beginning with this closure question, LES has several more fundamental questions that must be addressed before beginning any simulation. These include:

- How to close the filtered nonlinear term?
- What are the criteria for "success" in a given closure?
- How do the discrete equations communicate with molecular diffusion?
- What are the criteria for success in a given communication with molecular viscosity? If the microscale[4] of a continuum model is denoted by η_{model}, the answer to this question is at least known.

$$\eta_{\mathrm{model}} = \delta \Rightarrow Success,$$

$$\eta_{\mathrm{model}} << \delta \Rightarrow Failure.$$

- How to impose boundary conditions for the model? The choice here is either to accept a Re dependent mesh in the near wall region (often called "near wall resolution") or to try to retain an $O(\delta)$ meshwidth and Re independence of total computational complexity there (called "near wall modeling").
- How to discretize the given LES model on a given mesh so as to preserve the behavior the model was selected for?
- How to solve the given discrete equations efficiently?

1.3   Eddy Viscosity Closures

The most commonly used closure (which also addresses the question of how the model communicates with molecular viscosity) is by eddy viscosity models. Eddy viscosity models are treated at great length in many places

---

[4]The microscale η_{model} of a model is the length associated with the smallest persistent structure of solutions to a model. The microscale of the NSE is denoted η. $\eta = \eta_{NSE}$ is observed in flow data to be positive and in accord with estimates of them coming from turbulence phenomenology. Proving them to be so for weak solutions of the NSE would solve the Clay prize problem.

so herein we shall treat mostly the contribution approximate deconvolution ideas can make to improving eddy viscosity models. It is common in eddy viscosity modeling to rewrite the non closed, filtered nonlinear term as

$$\nabla \cdot (\overline{uu}) = \nabla \cdot (\overline{uu} - \overline{u}\ \overline{u}) + \nabla \cdot (\overline{u}\ \overline{u})$$
$$= \overline{u} \cdot \nabla \overline{u} + \nabla \cdot R(u, u), \text{ where } R(u, u) = \overline{uu} - \overline{u}\ \overline{u}.$$

Rewriting it this way the SFNSE becomes

$$\overline{u}_t + \overline{u} \cdot \nabla \overline{u} - \nu \triangle \overline{u} + \nabla \cdot R(u, u) + \nabla \overline{p} = \overline{f}(x, t), \text{ and } \nabla \cdot \overline{u} = 0,$$

and closure becomes the problem of replacing R with a tensor that depends only on $\overline{u}$, not u. The eddy viscosity hypothesis/Boussinesq approximation is the unresolved part R(u,u) has a dissipative effect on the mean flow. Mathematically this means

$$\nabla \cdot R(u, u) \simeq -\nabla \cdot (\nu_T \nabla^s \overline{u}) + \text{ terms incorporated into the pressure,}$$

where

$$\nabla^s \overline{u} := (\nabla \overline{u} + \nabla \overline{u}^{tr})/2,$$

$$\nu_T := \text{turbulent viscosity coefficient.}$$

The closure problem now becomes one of picking

$$\nu_T = \nu_T(\overline{u}, \delta, \text{Re}, \nu, \cdots).$$

The starting point in this search is the, so called, Kolmogorov–Prandtl relation. Motivated by the idea that the amount of turbulent mixing should depend on the kinetic energy in the turbulent fluctuations[5] and the local length scale of them, the simplest dimensionally consistent turbulent viscosity is

$$\nu_T = C\delta\sqrt{k'}, k' := \frac{1}{2}|u - \overline{u}|^2, \text{ which roughly means}$$

$$\nu_T = C\delta|u - \overline{u}|.$$

However the eddy viscosity is selected, the resulting model looks like the NSE with a nonlinear viscosity. Calling (w, q) the approximations to the

---

[5]This is variously denoted by TKE or k' (herein) or k. Again mathematics runs into the problem of not enough letters in the alphabet: k will also denote wave number in a Fourier series. Fortunately, there is very little chance in confusing wave number with TKE.

averaged velocity and turbulent pressure, we have

$$w_t + w \cdot \nabla w - \nabla \cdot \left(2\left[\nu + \nu_T(w, \delta)\right] \nabla^s w\right) + \nabla q = \overline{f}(x, t), \qquad (1.13)$$

$$\nabla \cdot w = 0 \qquad (1.14)$$

Eddy viscosity models differ based on how the turbulent kinetic energy is estimated. Eddy viscosity models work simply and wonderfully when the eddy viscosity amount and location is well tuned to the flow. They can fail by being under-diffused (and producing non-physical wiggles). They can also fail by being over-diffused and introducing too much model diffusion by overestimating the regions in which it is needed and overestimating the amount. It is our experience that since non-physical wiggles are easy to spot (and result in increased eddy viscosity in the next simulation), the most common failure mode is the latter. It is certainly widely reported for the Smagorinsky model (and has led to corrections such as van Driest damping). Three popular options are given by:

- The Smagorinsky model for which $k' \simeq \delta \frac{1}{2} |\nabla^s \overline{u}|$ and thus $\nu_T = (C_S \delta)^2 |\nabla^s \overline{u}|$.
- The conventional k-epsilon turbulence model, in which (non closed) equations are derived for k', involving another non closed equation for another variable ε. These are closed various ways and the total coupled system resulting is called the k-epsilon model.
- Approximate deconvolution methods of estimating k' such as $k' \simeq \frac{1}{2}|\overline{u} - \overline{\overline{u}}|$ and thus $\nu_T = C_{ADM} \delta |\overline{u} - \overline{\overline{u}}|$ and, more accurately and generally

$$\nu_T = C_{ADM} \delta |\overline{u} - D(\overline{u})|,$$

$$D := \text{approximate deconvolution operator.}$$

Eddy viscosity models have the tremendous advantage that if one has a code for laminar flows that is efficient and reliable then it can quickly be adapted to simulate turbulent flows based on eddy viscosity models. For example, if we suppress the secondary spacial discretization we can time step by

$$\frac{w^{n+1} - w^n}{\triangle t} + w^n \cdot \nabla w^{n+1} + \nabla q^{n+1}$$

$$-\nabla \cdot \left(2\left[\nu + \nu_T(w^n, \delta)\right] \nabla^s w^{n+1}\right) = \overline{f}(x, t_{n+1}),$$

$$\nabla \cdot w^{n+1} = 0.$$

Because the eddy viscosity coefficient $\nu_T(w^n, \delta)$ is non-negative, it can be treated explicitly and calculated from known values at previous time levels without altering numerical stability. This means that it adds very little programming or computational complexity to the starting point of a given program for laminar flows. This advantage is preserved for approximate deconvolution closures, treated throughout this book, that reduce the amount of model diffusion significantly.

1.4   Closure by van Cittert Approximate Deconvolution

The filter we will (mostly) use herein is a differential filter that satisfies

$$\overline{u} = u + O(\delta^2), \text{ for smooth } u.$$

An approximate filter inverse is a bounded linear operator D that provides an asymptotic inverse to the filter operator. For filters satisfying the above this typically means, for some $\beta \geq 2$

$$D(\overline{u}) = u + O(\delta^\beta), \text{ for smooth } u.$$

One can question the relevance of anything holding "for smooth u". When such an estimate is proven locally then it implies that high accuracy is attained in sub-regions where the flow is smooth (and then coarse meshes can be used there) and (in scale space) on the smooth flow components (and thus the larger and more energetically important scales are reconstructed with higher accuracy then the marginally resolved scales).

There are many examples of deconvolution operators included in software packages for image processing and filtering. Inversion of a smoothing filter is an ill posed problem so any method for solving ill posed problems (approximately) induces a deconvolution operator. At this point it is not yet clear which ones are the best for deconvolution of turbulent velocities. Several are reviewed in Sect. 3.2. However, in current practice in LES, the most common and highly developed is the family of van Cittert deconvolution operators, e.g., Bertero and Boccacci [BB98]. Its use in LES was pioneered by Stolz, Adams and Kleiser [AS01, SA99, AS02, SAK01a, SAK01b, SAK02].

Let the filter operator be denoted by G so

$$\overline{u} = Gu.$$

The Nth van Cittert approximate deconvolution operator D_N is defined by N steps of Picard iteration for the fixed point problem:

given $\overline{u}$ solve $u = u + \{\overline{u} - Gu\}$ for u.

The van Cittert approximate deconvolution operator D_N is defined as follows.

Algorithm 5 (van Cittert Approximate deconvolution). *Set* $u_0 = \overline{u}$
    For $n = 1, 2, \ldots, N - 1$, *perform*
$$u_{n+1} = u_n + \{\overline{u} - Gu_n\}$$
    Define $D_N \overline{u} := u_N$.

By eliminating the intermediate steps, the Nth deconvolution operator D_N is given explicitly by

$$D_N \phi := \sum_{n=0}^{N} (I - G)^n \phi. \tag{1.15}$$

The van Cittert approximate deconvolution operator corresponding to $N = 0, 1, 2$ and their formal orders of accuracy are:

$$D_0 \overline{u} = \overline{u} = u + O(\delta^2),$$
$$D_1 \overline{u} = 2\overline{u} - \overline{\overline{u}} = u + O(\delta^4),$$
$$D_2 \overline{u} = 3\overline{u} - 3\overline{\overline{u}} + \overline{\overline{\overline{u}}} = u + O(\delta^6).$$

Approximate deconvolution in LES has several constraints not binding in image processing. Aside from accuracy and stability of the induced model, the method used must be fast, parallel and not require significant extra storage. The van Cittert algorithm requires repeated filtering. When filtering is done by a local averaging, van Cittert is very appealing for the needs of LES. It is an open problem to test advanced deconvolution methods from image processing in LES.

Since $D(\overline{u}) \simeq u$, an approximate deconvolution operator gives an immediate and systematic solution to the closure problem by

$$\nabla \cdot (\overline{u\,u}) = \nabla \cdot \left(\overline{D(\overline{u})\ D(\overline{u})} \right) + O(\delta^\beta)\,, \quad \text{for smooth } u.$$

This yields the *basic* approximate deconvolution LES model

$$w_t + \overline{D(w) \cdot \nabla D(w)} - \nu \triangle w + \nabla q = \overline{f}(x, t), \tag{1.16}$$
$$\nabla \cdot w = 0. \tag{1.17}$$

The simplest case is $D = I$ which gives the zeroth order model, given by

$$w_t + \overline{w \cdot \nabla w} - \nu \triangle w + \nabla q = \overline{f}(x, t), \tag{1.18}$$
$$\nabla \cdot w = 0. \tag{1.19}$$

There is a strong connection between the zeroth order model and the general ADM that explains much of the correspondence of the analysis in the general case with the zeroth order case. To see this connection, define a new filter by

$$\widetilde{u} := D(\overline{u}).$$

Applying the deconvolution operator D to the ADM (1.16) and defining the new filtered approximation to u, $v := D(w)$, shows that v satisfies

$$v_t + \widetilde{v \cdot \nabla v} - \nu \triangle v + \nabla q = \widetilde{f}(x, t), \qquad (1.20)$$

$$\nabla \cdot v = 0, \qquad (1.21)$$

which is just the zeroth order model (1.18) with a different filter.

If the filter is a differential filter $\overline{u} = (-\delta^2 \triangle + 1)^{-1} u$, then the ADM can be understood by applying $-\delta^2 \triangle + 1$ to the equations. This shows that the ADM is equivalent to

$$w_t - \delta^2 \triangle w_t + D(w) \cdot \nabla D(w) - \nu \triangle w + \delta^2 \triangle^2 w + \nabla q = \overline{f}(x, t), \qquad (1.22)$$

$$\nabla \cdot w = 0. \qquad (1.23)$$

For the simplest case of the zeroth order model we have the *fourth order* problem

$$w_t - \delta^2 \triangle w_t + w \cdot \nabla w - \nu \triangle w + \delta^2 \triangle^2 w + \nabla q = \overline{f}(x, t), \qquad (1.24)$$

$$\nabla \cdot w = 0. \qquad (1.25)$$

Thus, the ADM introduces two effects even in the simplest case:

1. An extra hyperviscosity term in the term $\delta^2 \triangle^2 w$.
2. A modification of the model's kinetic energy in the term $\delta^2 \triangle w_t$.

Since it is easy to generate approximate deconvolution operators that are of very high order accuracy, ADMs give a general and accurate solution of the closure problem. The main challenges of ADMs are:

- Using the formulation (1.22) requires the solution of either a fourth order problem or two coupled second order problems. Both are known to increase the complexity of solving the model.
- Using the formulation (1.16) requires treating the full nonlinear term explicitly. This is often done in CFD codes but does introduce a CFL condition on the time step.
- Boundary conditions: fourth order problems require extra boundary conditions. Since $-\delta^2 \triangle \overline{u} + \overline{u} = u$, it is natural (but not well tested) to select the extra boundary conditions to be

$$w = 0 \text{ and } \triangle w = 0 \text{ on walls}.$$

- The base ADM does not include sufficient numerical diffusion by itself to move the model microscale to $O(\delta)$. We shall see in Chap. 3 and Sect. 1.5

how to accomplish this within the ADM framework by adding a time relaxation term.

Chapter 4 shows how the phenomenology of ADMs of turbulence is derived. It shows that ADMs produce the correct statistics of homogeneous, isotropic turbulence. We show

$$E(w)(k) \simeq \alpha_{model}\varepsilon_{model}^{2/3}k^{-5/3}, \text{ for } k \leq \frac{1}{\delta},$$

$$E(w)(k) \simeq \alpha_{model}\varepsilon_{model}^{2/3}\delta^{-2}k^{-11/3}, \text{ for } k \geq \frac{1}{\delta}.$$

Thus, *above the cutoff length scale the ADM predicts the correct energy cascade!* Without time relaxation (defined in the next section), the model microscale is

$$\eta_{model} \simeq Re^{-\frac{1}{3}}L^{\frac{4}{9}}\frac{\delta^{\frac{2}{3}}}{(1+(\frac{\delta}{L})^2)^{\frac{1}{9}}}(>> \eta_{NSE}).$$

With an additional time relaxation term, if U, L, N denote the global velocity scale, length scale and order of van Cittert deconvolution, choosing the time relaxation coefficient's value to be

$$\chi_{optimal} \simeq \frac{U}{L^{\frac{1}{3}}}2^{N+1}\delta^{-\frac{2}{3}} \tag{1.26}$$

gives microscale equal to the filter length

$$\eta_{model} \simeq \delta.$$

Thus, time relaxation is an important regularization in itself and as an addition to models which have good properties otherwise but fail the essential test that the microscale be comparable to the filter width. Testing the dependency of $\chi_{optimal}$ upon N would require a spectral code since, as N increases, deconvolution errors are soon much smaller than numerical errors. The dependence on N is possibly related to the theoretical prediction that as N increases δ should be (slowly) increased to fix the induced cutoff to a preselected value.

1.4.1 The Bardina Model

The Bardina model [Bar83] is a different approach to LES modeling than considered herein. It is given by

$$w_t + w \cdot \nabla w + \nabla \cdot (\overline{ww} - \overline{w}\,\overline{w}) - \nu\triangle w + \nabla q = \overline{f}(x,t), \tag{1.27}$$

$$\nabla \cdot w = 0. \tag{1.28}$$

Deconvolution can also be used to increase the accurate of scale similarity/ Bardina type (e.g., [Bar83]) models. However, there is no proof of their stability and computations with the $N = 0$ case (the Bardina model) without extra eddy viscosity have hinted at stability problems.

1.4.2 The Accuracy of van Cittert Deconvolution

Computational tests of deconvolution models and regularizations have been unequivocal: higher formal accuracy deconvolution (up to the limit imposed by the inherent errors in the numerical method used) give results that are both quantitatively and qualitatively superior. This gives a fundamental challenge for analysis of models: explain the connection between formal accuracy (for smooth velocities) in deconvolution that drives greater accuracy in computed solutions to ADM based models of flows whose solutions are not smooth. This analytical problem has many unknowns. To explain the issues, we begin by noting that the accuracy of van Cittert deconvolution can be measured in different ways.

1.4.2.1 The Accuracy for Smooth Functions

Since van Cittert deconvolution is mathematically equivalent to a truncation of a geometric (operator) series, it is quite easy to calculate the deconvolution error for specific choices of filter for smooth functions. For example, suppose $\overline{u} = Gu = (-\delta^2 \triangle + 1)^{-1} u$. Then we have (see Lemma 18 of Chap. 3)

$$u - D_N \overline{u} = (I - G)^{N+1} u = (-1)^{N+1} \delta^{2N+2} \triangle^{N+1} G^{N+1} u$$

$$= O(\delta^{2N+2}) \text{ for } C^{\infty}_{periodic} \text{ functions } u.$$

This is the most optimistic case and it does affirm an important feature of van Cittert based ADMs: that they are consistent with the NSE to a high level on the smooth flow components.

1.4.2.2 The Accuracy in the Limit $Re \to \infty$

On the other hand, turbulent velocities are not $C^{\infty}_{periodic}$ functions even far away from walls. Some insight into worst case accuracy in the limit as $Re \to \infty$ was obtained in [LL05] as follows. If we postulate a time averaged energy spectrum that is consistent with the K41 theory that

$$E(k) \leq \alpha \varepsilon^{2/3} k^{-5/3},$$

then the time averaged deconvolution error can be calculated directly by Fourier methods.

Theorem 6. *Suppose $E(k) \leq \alpha\varepsilon^{2/3}k^{-5/3}$ then*

$$< ||u - D\overline{u}||^2 >^{1/2} \leq C(N)\sqrt{\alpha}\varepsilon^{1/3}\delta^{1/3}.$$

Proof. Splitting the integral below depending on which term in the denominator dominates, we calculate

$$< ||u - D\overline{u}||^2 > = \sum_{k \geq 1} |1 - \widehat{D}(k)\widehat{G}(k)|^2 E(k)$$

which is majorized by the integral

$$\leq \int_0^\infty |1 - \widehat{D}(k)\widehat{G}(k)|^2 E(k) dk$$

$$= \int_0^\infty |\frac{(\delta k)^2}{(\delta k)^2 + 1}|^{2N+2} E(k) dk$$

$$\leq \left(\int_0^{k\delta=1} |(\delta k)^2|^{2N+2} \alpha\varepsilon^{2/3}k^{-5/3} dk + \int_{k\delta=1}^\infty \alpha\varepsilon^{2/3}k^{-5/3} dk \right)$$

$$\leq \int_0^{k\delta=1} |(\delta k)^2|^{2N+2} \alpha\varepsilon^{2/3}k^{-5/3} dk$$

$$+ \int_{k\delta=1}^\infty \alpha\varepsilon^{2/3}k^{-5/3} dk \leq C(N)\alpha\varepsilon^{2/3}\delta^{2/3},$$

uniformly in the Reynolds number. Thus we conclude that, uniform in Re,

$$< ||u - D\overline{u}||^2 >^{1/2} \leq C(N)\sqrt{\alpha}\varepsilon^{1/3}\delta^{1/3}$$

$$\square$$

This time averaged rate of $O(\delta^{1/3})$ is not encouraging and does not fit with computational experience. However, we note that it is the pessimistic worst case rate holding in the limit of infinite Re and for *all scales* both resolved and unresolved. Looking carefully at the above calculation we find:

- If we evaluate the deconvolution error over the resolved scales alone, $0 < k \leq 1/\delta$, it is again only $O(\delta^{1/3})$.
- If, however, we fix a range of scales, $0 < k \leq k_{\max}$, and only evaluate the error over this fixed range of scales we recalculate and find that accuracy is governed by the smooth case.

1.4.2.3 The Accuracy in Statistics

It is an open question to understand the right answer (or even if there is one right answer). The right answer must change according to what statistic one seeks. If the statistic (number) is the result of a linear functional $u \to I(u)$, it has the form:

$$I(u) = < \int_\Omega \phi \cdot u \, dx > \quad \text{for } \phi(x) \text{ a fixed function.}$$

The error in calculating the statistic with $D_N \overline{u}$ is then

$$Error = < \int_\Omega \phi \cdot (u - D_N \overline{u}) \, dx > .$$

If the statistic is sensitive to small scales then the error is likely to be determined by the pessimistic case by (using Cauchy-Schwarz)

$$|Error| \leq ||\phi|| \; < ||u - D_N \overline{u}||^2 >^{1/2} \leq C(N, ||\phi||, \varepsilon) \delta^{1/3}.$$

On the other hard, if the statistics sought depend mostly on the energy in the large scales then ϕ is smooth and we may see a higher rate of convergence because $|Error| \leq ||\phi||_{H^s} \; < ||u - D_N \overline{u}||^2_{H^{-s}} >^{1/2}$, where the norm is defined using Fourier series by

$$||u||^2_{H^{-s}} := \sum_{k \geq 1} (1 + k^2)^{-s} \sum_{|\mathbf{k}|=k} |\widehat{u}(\mathbf{k})|^2.$$

Theorem 7. *Suppose* $E(k) \leq \alpha \varepsilon^{2/3} k^{-5/3}$ *then*

$$< ||u - D\overline{u}||^2_{H^{-s}} >^{1/2} \leq C(N, s) \sqrt{\alpha} \varepsilon^{1/3} \delta^{1/3+s}, \quad over \; 0 \leq s \leq 2N + 2 - 7/6.$$

Proof. First note that the norm is equivalent to the one easier to calculate given by

$$||u||^2_{H^{-s}} \simeq \sum_{k \geq 1} k^{-2s} E(k).$$

Thus, following the proof of the last theorem, we calculate

$$< ||u - D\overline{u}||^2_{H^{-s}} >$$

$$\leq \sum_{k \geq 1} |1 - \widehat{D}(k)\widehat{G}(k)|^2 k^{-2s} E(k)$$

$$\leq \int_0^\infty |1 - \widehat{D}(k)\widehat{G}(k)|^2 k^{-2s} E(k) dk$$

$$= \int_0^\infty |\frac{(\delta k)^2}{(\delta k)^2 + 1}|^{2N+2} k^{-2s} E(k) dk$$

$$\leq \left(\int_0^{k\delta=1} |(\delta k)^2|^{2N+2} \alpha \varepsilon^{2/3} k^{-5/3-2s} dk + \int_{k\delta=1}^\infty \alpha \varepsilon^{2/3} k^{-5/3-2s} dk \right)$$

$$\leq \int_0^{k\delta=1} |(\delta k)^2|^{2N+2} \alpha \varepsilon^{2/3} k^{-5/3-2s} dk + \int_{k\delta=1}^\infty \alpha \varepsilon^{2/3} k^{-5/3-2s} dk$$

$$\leq C(N,s)\alpha \varepsilon^{2/3} \delta^{2/3+2s}, \text{ over } 0 \leq s \leq 2N + 2 - 7/6. \qquad \square$$

For smooth ϕ we conclude *the error in approximation of statistics depending on the largest energetic scales by deconvolution approximation is much better:* up to $O(\delta^{2N+7/6})$:

$$|Error| \leq ||\phi||_{H^s} < ||u - D_N \overline{u}||_{H^{-s}}^2 >^{1/2} \leq C(N, ||\phi||_{H^s}, \varepsilon)\delta^{2N+7/6}.$$

Thus, for calculating quantities depending on the largest energetic scales, the smooth case errors *may* be attained. The next challenge for analysts is to prove negative norm estimates for the error in ADMs- something not yet accomplished for usual numerical methods for approximating laminar flows of the NSE!

1.4.2.4   The Accuracy as $N \to \infty$ for Fixed δ

There is very limited analysis of the behavior as $N \to \infty$ for fixed δ and many open problems. At first sight, this whole question seems to emerge from mathematicians perverse instincts to let $N \to \infty$ "because it is there". Nevertheless, the limit can be sensible since the computational cost of cutting $\delta \to \delta/2$ is to remesh with $8x$ per points per time step. Fixing δ and increasing $N \to N + 1$ requires only one extra filtering step on a fixed mesh per time step. Relatively little is known about this limit. In [LL08] it was proven that the solution of the Leray deconvolution model converges to a weak solution of the NSE as $N \to \infty$ for fixed δ. This was recently extended by Berselli and Lewandowski [BL11] to the harder case of the full ADM. Interestingly, neither result is completely positive for practical LES with large N. The explanation seems to be that (as pointed out in [LN06b]) as $N \to \infty$ for fixed δ the effective cutoff length scale decreases to zero as well. Thus the interesting limit (and still open for analysis) is to show how $N \to \infty$ and in a related manner δ *increases* so that the model solution approaches the NSE solution *projected upon a fixed set of scales* (related to computational resources)! If this conjecture is correct, then the computational cost may actually decrease slowly as $N \to \infty$.

The open questions arising in understanding the error in LES ADMs are many and subtle!

1.5 Approximate Deconvolution Regularizations

Because of the intricacies of solving a full approximate deconvolution model of turbulence and the difficulties with commutator errors and near wall models, there has been considerable interest in exploring turbulent flow simulations based on simpler regularizations of the NSE rather than full LES models. NSE regularizations can be simpler for numerical simulation than a full turbulence model and have mathematical properties more favorable for numerical solution than the unregularized NSE. However, the computed solution is then simply a regularized approximation of the NSE solution rather than a local, spacial average of the fluid velocity. Chapters 6 and 7 study approximate deconvolution regularization modeling.

The many possible NSE regularizations can be judged based on (a) accuracy *as an approximation of the NSE*, (b) truncation of scales and (c) fidelity to qualitative properties of the NSE's velocity. Fundamentally, a successful regularization should:

- Have unique smooth, strong solutions
- Have a lucid energy balance
- Have small consistency error in smooth regions, for laminar flows and on the flow's large scales
- Be simple for numerical solution
- Impose a microscale $\eta_{\text{model}} = O(\delta)$

Beyond those basic criteria, there is a great need to develop *quantifiable* criteria for assessing and comparing the *predictive skill* of NSE regularizations. Without quantitative comparisons the interesting approach of NSE regularization is in danger of splintering into a myriad of disconnected regularizations for which the same proofs are repeated and the essential problems are avoided.

1.5.1 Time Relaxation

The simplest (and perhaps the best) regularization and one which is a fundamental component to many more complex models is time relaxation.

Time relaxation adds to the NSE one linear or nonlinear term which adds extra model diffusivity at the cutoff length scale. In the case of linear time relaxation, we consider

$$w_t + w \cdot \nabla w - \nu \triangle w + \nabla q + \chi(w - D(\overline{w})) = f(x,t), \qquad (1.29)$$

$$\nabla \cdot w = 0.$$

Time relaxation is related to Newtonian damping and to "nudging" in data assimilation; the extra term acts to nudge the flow to its own large scale components. Discretization of time relaxation terms is very simple: it can be lagged without altering stability by, for example,

$$\frac{w^{n+1} - w^n}{\triangle t} + w^n \cdot \nabla w^{n+1} + \nabla q^{n+1} - \nu \triangle w^{n+1}$$

$$+\chi(w^{n+1} - D(\overline{w^n})) = f(x, t_{n+1}),$$

$$\nabla \cdot w^{n+1} = 0.$$

The main point of time relaxation is that it can force a good model microscale with minimal effect on the resolved scales. We show in Chap. 5 that with D the Nth van Cittert deconvolution operator, choosing time relaxation coefficient's value to be

$$\chi_{optimal} \simeq \frac{U}{L^{\frac{1}{3}}} 2^{N+1} \delta^{-\frac{2}{3}} \qquad (1.30)$$

gives the microscale predicted by turbulence phenomenology to be equal to the filter length

$$\eta_{model} \simeq \delta.$$

With $\chi_{optimal} \simeq O(\delta^{-2/3})$ the *consistency error* (in the smooth regions) induced by the added regularization term is

$$\text{Consistency error} = ||\chi(u - D(\overline{u}))|| = O(\delta^{2N+4/3}).$$

This choice $\chi_{optimal} \simeq O(\delta^{-2/3})$ is deceptively satisfying; it is derived under the full assumptions of K41 phenomenology of homogeneous, isotropic turbulence. When these assumptions are violated, such as in boundary layers, different choices are likely necessary and corresponding formulas are not yet known. It seems that choosing different values of the relaxation parameter χ at different meshpoints and even nonlinear time relaxation may both be needed for real flows.

1.5.2 The Leray-Deconvolution Regularization

In 1934 J. Leray [L34a, L34b] proved that the *regularized* Navier–Stokes equations has a unique, smooth, strong solution and that as the regularization length-scale $\delta \to 0$, the regularized system's solution converges (modulo a subsequence) to a weak solution of the Navier–Stokes equations. If $\overline{w}$ denotes a local spacial average of the velocity w associated with filter length-scale δ, the classical Leray model is given by

$$\frac{\partial w}{\partial t} + \overline{w} \cdot \nabla w - \nu \triangle w + \nabla q = f, \tag{1.31}$$

$$\nabla \cdot w = 0. \tag{1.32}$$

In retrospect it is a simple step to increase accuracy by introducing deconvolution. At the time, Adrian Dunca's idea to do so was a significant step forward that set the pattern for many similar improvements of other regularizations. Dunca's Leray-type deconvolution model is

$$w_t + D(\overline{w}) \cdot \nabla w - \nu \triangle w + \nabla q = \overline{f}, \ and \ \nabla \cdot w = 0.$$

The Leray-deconvolution models have arbitrarily high orders of accuracy and include the Leray model as the zeroth order ($N = 0$) case. The Leray deconvolution model is developed in Chap. 6. It has the following attractive properties:

- For $N = 0$ they include the Leray/Leray-α model as the lowest order special case.
- Their accuracy is high, $O(\delta^{2N+2})$ for arbitrary $N = 0, 1, 2, \ldots$.
- They improve upon the attractive theoretical properties of the Leray model, e.g. convergence (modulo a subsequence) as $\delta \to 0$ to a weak solution of the NSE and $||u_{NSE} - u_{LerayDCM}|| = O(\delta^{2N+2})$ for a smooth, strong solution u_{NSE}.
- Given u the computation of $D_N \overline{u}$ is computationally attractive.
- The higher order models (for $N \geq 1$) give dramatic improvement of accuracy and physical fidelity over the $N = 0$ case.
- Increasing model accuracy can be done in two ways: (a) cutting $\delta \to \delta/2$ increases accuracy for $N = 0$ by approximately a factor of 4 but requires remeshing with approximately 8 times as many unknowns, and (b) increasing $N \to N + 1$ increases accuracy from $O(\delta^{2N+2})$ to $O(\delta^{2N+4})$ and requires one more Poisson solve $((-\delta^2 \triangle + 1)^{-1} \phi)$ per time step.

1.5.3   The NS-Alpha Regularization

The NS-α model is a recently developed regularization of the NSE with desirable mathematical properties, pioneered by work of Holm, Foias, Titi and their collaborators. The literature on the NS-α regularization is large and growing, see, e.g., [FHT01, FHT02, GOP03] as beginning points. It is given by

$$w_t - \overline{w} \times (\nabla \times w) + \nabla Q - \nu \Delta w = f, \text{ in } \Omega \times (0, T], \tag{1.33}$$

$$\nabla \cdot \overline{w} = 0, \text{ in } \Omega \times (0, T], \tag{1.34}$$

$$\overline{w} = G(w), \text{ in } \Omega \times (0, T], \tag{1.35}$$

$$w(x, 0) = u_0(x), \text{ in } \Omega. \tag{1.36}$$

There are very many approaches to physical and numerical BCs in CFD. It does seem plausible however that the place to begin is with periodic and no slip BCs. Thus we shall impose

$$w = 0, \text{ on } \partial\Omega.$$

Since the averaged term $\overline{w}$ in the nonlinearity is nonlocal, any reference to it in the discrete system that is implicit must be implemented as a coupled system for $(w, \overline{w}, Q)$. However, in finite element spaces with N_V velocity degrees of freedom and N_P pressure degrees of freedom, *the method leads to a large, nonlinear system at each time step with $2N_V + N_P$ total degrees of freedom.* This is the first computational challenge in solving the NS-alpha model: many extra variables must be introduced when using implicit methods. To avoid this, the only alternatives are:

1. Use a different regularization such as Leray or NS-omega (next).
2. Use explicit methods for the NS-alpha nonlinearity. The second computational challenge is associated with the use of the rotational form of the NSE nonlinearity in under resolved flow simulations.

   Stability of this fully coupled discretization is proven in Chap. 7.

1.5.4   The NS-Omega Regularization

A motivation for the modification from "alpha" to "omega" is the search for efficient, unconditionally stable and (at least second order) accurate methods for the simulation of under-resolved flows. The NS-ω model is derived from the rotational form of the NSE by filtering the second term of the nonlinearity

$$w_t - w \times (\nabla \times \overline{w}) + \nabla q - \nu \Delta w = f, \tag{1.37}$$

$$\nabla \cdot w = 0, \tag{1.38}$$

$$-\delta^2 \Delta \overline{w} + \overline{w} = w. \tag{1.39}$$

This simple (and in retrospect easy to see) modification yields a model with stronger stability properties which is amenable to numerical simulations. Comparing the alpha model to the omega model, the NS-ω model averages the vorticity term $\omega = \nabla \times u$ rather than the velocity term in the nonlinearity. (Hence calling it the NS-ω regularizations seems descriptive.) The structure of the nonlinearity in the NS-ω model admits simple methods which are nonlinearly, unconditionally stable, second order accurate, *linearly implicit* (only 1 linear system per time step) and require only $N_V + N_P$ total degrees of freedom. The following attractive variant CNLE (Crank-Nicolson with linear extrapolation, known for the NSE at least since Baker's 1976 paper [B76]) is one possibility. Let $w^{n+1/2} := (w^{n+1} + w^n)/2$ and $U^n := \frac{3}{2}w^n - \frac{1}{2}w^{n-1}$. Then the calculation of $\overline{U^n}$ and $D(\overline{U^n})$ are an explicit and uncoupled calculation of filtering and deconvolution of a known (from previous time values) function. Thus, the following requires no extra storage and only minimal extra operations over the same method for the NSE:

$$\frac{w^{n+1} - w^n}{\Delta t} + \nabla \times D(\overline{U^n}) \times w^{n+\frac{1}{2}} - \nu \Delta w^{n+\frac{1}{2}}$$

$$+ \nabla Q^{n+\frac{1}{2}} + \chi(w^{n+1/2} - D(\overline{w^n})) = f^{n+\frac{1}{2}},$$

$$\nabla \cdot w^{n+\frac{1}{2}} = 0.$$

It is unconditionally stable (take the inner product with $w^{n+\frac{1}{2}}$). It is clearly second order accurate and linearly implicit (since $U^n := \frac{3}{2}w^n - \frac{1}{2}w^{n-1}$ is known from previous time levels). Further, since $U^n := \frac{3}{2}w^n - \frac{1}{2}w^{n-1}$ is known, its average can be directly computed, uncoupled from the linear equations for advancing in time.

1.6   The Problem of Boundary Conditions

If $\overline{u}$ is an average of u of averaging radius δ, then $\overline{u}$ can be well approximated on a mesh with $\Delta x = \Delta y = \Delta z = \delta$, and thus with cost independent of the Reynolds number. This is the simultaneously plausible and astonishing claim of LES. The first sticking point is interior closure: finding equations that predict $\overline{u}$ to high accuracy. The second and more difficult sticking point is the problem of finding boundary conditions for these equations for $\overline{u}$. While this presentation is about the interior model, a fair assessment requires a description of the problem of boundary conditions. To illustrate, consider the problem of boundary conditions for $\overline{u}$ at a fixed solid wall where the no slip condition for u holds. There are several difficult issues.

1.6.1 The Commutator Error

There are considerable technical details, but averaging the NSE in the presence of walls by a convolution filter $(u \to g_\delta \star u = \overline{u})$ reveals that in the presence of walls an extra commutator error term arises in the SFNSE

$$\overline{u}_t + \nabla \cdot (\overline{u}\,\overline{u}) + \nabla \overline{p} - \nu \Delta \overline{u} + A_\delta(u, p) = \overline{f}, \text{ and } \nabla \cdot \overline{u} = 0, \qquad (1.40)$$

$$A_\delta(u, p) = \text{ commutator error term.} \qquad (1.41)$$

See [DJL04,DM01,TS06,BGJ07,LT10] for its derivation, precise specification and some analysis. Sadly,

$$||A_\delta(u, p)||_{L^2} \to \infty \text{ as } \delta \to 0$$

because it piles up to infinity in the near wall region as $\delta \to 0$ so this extra commutator error term is not negligible. While there has been some analysis of its behavior, there has only been one proposal of Das and Moser for its modeling. Relatively simple models, such as $A_\delta(u, p) \Leftarrow A_\delta(\overline{u}, \overline{p})$, have neither been analyzed nor tested.

1.6.2 Near Wall Modeling

Even with an accurate commutator closure for $A_\delta(u, p)$ for the near wall region, the problem remains of finding boundary conditions for $\overline{u}$. Essentially $\overline{u}|_{walls}$ depends non-locally on the behavior of $u(x, t)$ *near* the wall. Thus, *boundary conditions for averages on walls can be no better than our (poor) understanding of turbulent boundary layers themselves.* We give a simple and fairly typical attempt to generate boundary conditions (near wall laws) of the form originally proposed by Navier for the NSE:

$$\begin{cases} \overline{u} \cdot \hat{n} = 0, \text{ on } \partial\Omega, \text{ and} \\ \overline{u} \cdot \hat{\tau}_j + \beta(\delta, Re)\hat{n} \cdot \mathbf{\Pi}(\overline{u}, \overline{p}) \cdot \hat{\tau}_j = 0 \text{ on } \partial\Omega, \end{cases} \qquad (1.42)$$

where $(\hat{n}, \hat{\tau}_j)$ are the unit tangent and normal vectors and $\mathbf{\Pi}(\overline{u})$ is the stress tensor associated with the viscous and turbulent stresses:

$$\hat{n} \cdot \mathbf{\Pi} \cdot \hat{\tau}_j = \hat{n} \cdot (2\nu\nabla^s\overline{u} + \nu_T\nabla^s\overline{u}) \cdot \hat{\tau}_j.$$

The friction coefficient $\beta = -\overline{u} \cdot \hat{\tau}_j/\hat{n} \cdot \mathbf{\Pi}(\overline{u}) \cdot \hat{\tau}_j$ is calculated replacing $\overline{u}$ by an average of a boundary layer approximation to u. One then typically gets relations like $\beta \sim Re\delta/L$ so as $\delta \to 0$ we recover the no-slip condition and as $Re \to \infty$ we transition to free-slip and the Euler equations. Clearly the resulting BC can be no better than the accuracy of the boundary layer approximation used to calculate the friction coefficient!

Much more sophisticated near wall laws built essentially on the same ideas are surveyed in the very interesting review articles [P08, PB02]. One conclusion seems to be that there is no current near wall model that is entirely satisfactory. This may reflect that models of the commutator error are usually omitted in derivation of near wall models or that our current understand of near wall turbulence needs to advance.

1.6.3 Changing the Averaging Operator to a Differential Filter

One attractive idea of Germano is to change the averaging operator and let the (implicitly induced) Green's function adapt the effective averaging radius to the near wall region and possible complicated geometry. For example, averages can be defined as the solution to

$$-\delta^2 \triangle \overline{u} + \overline{u} = u, \text{ in } \Omega \text{ and } \overline{u} = 0 \text{ on } \partial\Omega,$$

or, so as to preserve incompressibility, by a shifted Stokes problem

$$-\delta^2 \triangle \overline{u} + \overline{u} + \nabla\lambda = u, \text{ and } \nabla \cdot \overline{u} = 0 \text{ in } \Omega$$

$$\overline{u} = 0 \text{ on } \partial\Omega.$$

In either case the issue with boundary conditions is hidden. For example, for the first option since

$$-\delta^2 \triangle \overline{u}(x) + \overline{u}(x) = u(x)$$

$$\text{and as } x \to \partial\Omega : \overline{u}(x) \to 0, \text{ and } u(x) \to 0$$

we have the implicitly imposed boundary condition that

$$\triangle \overline{u} = 0 \text{ on } \partial\Omega.$$

For typical profiles this implies that $\overline{u}$ is *nearly linear in the near wall region,* which is hardly the correct physical behavior. Thus the boundary condition $\overline{u} = 0$ on $\partial\Omega$ seems wrong for differential filters. We are led back to imposing the BC (1.42) on the differentially filtered velocity and back to the reliability of boundary layer based calculation of friction coefficients.

Differential filters shift the problem of the commutator error to one which, while simpler, is still open. To see this, write the NSE as

$$u_t + u \cdot \nabla u + \nabla \cdot \Pi = f, \quad \nabla \cdot u = 0. \tag{1.43}$$

Filtering the NSE we must compute the filtered stress tensor $\overline{\Pi}$ or, more precisely, $\overline{\nabla \cdot \Pi}$. With the first differential filter we impose $\overline{\nabla \cdot \Pi} = 0$ on $\partial\Omega$ and with the second we additionally impose $\nabla \cdot \left(\overline{\nabla \cdot \Pi} \right) = 0$. Since neither makes sense, the only reasonably interpretation of differentially filtered NSE seems to be as the fourth order problem for $\overline{u}$:

$$\overline{u}_t - \delta^2 \triangle \overline{u} + u \cdot \nabla u - \nu \triangle (\overline{u} - \delta^2 \triangle \overline{u}) = f, \quad \nabla \cdot \overline{u} = 0. \tag{1.44}$$

Thus differential filters solve the commutator error problem but increase the order of the model and thus require extra boundary conditions.

1.6.4   Ad Hoc Corrections and Regularization Models

Regularization models are discussed above. Ad hoc corrections to the interior closure models themselves, such as van Driest damping most famously, can be successful for problems for which there is extensive computational experience.

1.6.5   Near Wall Resolution

It is entirely possible that there simply is no general "solution" beyond using an accurate model away from walls on coarser meshes and then using fine meshes in the near wall region. This is connected to taking the averaging radius $\delta = \delta(x) \to 0$ as $x \to walls$ and the mesh is refined accordingly. There is some experience with near wall resolution strategies and the design of filters with variable averaging radii to ameliorate the (new) commutator errors that then occur. While this results in complexity depending (weakly) on the Reynolds number, NWR, taking advantage of reduced storage and costs from constant δ away from walls, could well be the best chance at a general purpose strategy.

1.7   Ten Open Problems in the Analysis of ADMs

We close this preface with a list of ten open analytical problems. The "fundamental problem" in the theory of the Navier–Stokes equations, the million dollar Clay prize problem, necessarily pervades any analytical treatment of turbulence. The problem statement given by the Clay foundation has the benefits of brevity and clarity. For LES, the modification would read: *Find an LES model parameterized by the filter scale δ such that model selects a weak or strong solution of the NSE as its unique limit as $\delta \to 0$. The solution of* the fundamental problem in any form would be a giant advance in completing our understanding of turbulence and connecting it to our still rudimentary understanding of the Navier–Stokes equations. Here are nine more problems that are closer to our present analytical techniques:

1. Expand ADM phenomenology to give insight into the accuracy of model predictions of other aspects of turbulence, such as intermittence and turbulence modulated by other effects, such as MHD turbulence, rotation and temperature effects.
2. Elaborate the connection between a model's formal consistency error and the accuracy of model predictions. This requires developing estimates of model errors in negative Sobolev norms.
3. Develop a theory of turbulent boundary layers predicted by ADMs and modified by time relaxation. Use that theory to optimize deconvolution and time relaxation in the near wall region.
4. Expand the analytical foundations of the Near Wall Models, surveyed in Piomelli and Balaras [PB02, P08], used in practical computations.
5. Develop accurate models of the commutator error that are stable when incorporated into ADMs. Perform careful tests using them with simple near wall models to study how much of the difficulties reported in the near wall region are due to commutator errors and how much to errors in near wall models.
6. Extend the analytical foundations of ADMs to filtering by local averaging operators (for which $\widehat{g}_\delta(k) \not> 0$) and to better deconvolution operators.
7. Expand the theory of ADMs to include variable filter widths $\delta(x) \to 0$ as $x \to \partial\Omega$.
8. Develop ways to synthesize different models smoothly to combine the advantages of diverse models and retain a strong analytic foundation for the combination.
9. Extend the mathematical foundations of ADMs to compressible flows.

Chapter 2
Large Eddy Simulation

2.1 The Idea of Large Eddy Simulation

"The distinguishing feature of a turbulent flow is that its velocity field appears to be random and varies unpredictably. The flow does, however, satisfy the Navier–Stokes differential equations, which are not random. This contrast is the source of much of what is interesting in turbulence theory....

....One should keep in mind that a practical person is usually interested only in mean properties of a small number of functionals of the flow (e.g., lift and drag in the case of flow past a wing), and these could conceivably be obtained even when the details of the flow are unknown..."

A.J. Chorin, in: *Lectures on Turbulence Theory*, Publish or Perish, 1975.

"The upper limit to the size of an eddy is, like the length of a piece of string, a matter of human convenience."

L.F. Richardson, 1922 (p. 65 in [R22])

Models developed for Large Eddy Simulation (LES) (and especially eddy-viscosity models) are motivated by two physical ideas:

2.1.1 *Differing Dynamics of the Large and Small Eddies*

At high Reynolds number the fluid velocity is exponentially sensitive to perturbations of the problem data. This sensitivity, however, is not uniform. The large structures (large eddies) evolve deterministically and are thus not sensitive [BFG02]. The small eddies are sensitive because they have a random character. Their random character does, however, have universal features so that there is hope that their mean effects on the large eddies can be modeled. This idea was described already in 1510 by Leonardo da Vinci:

"Observe the motion of the water surface, which resembles that of hair, that has two motions: One is due to the weight of the shaft, the other due to the shape of

W.J. Layton and L.G. Rebholz, *Approximate Deconvolution Models of Turbulence*, Lecture Notes in Mathematics 2042, DOI 10.1007/978-3-642-24409-4_2, © Springer-Verlag Berlin Heidelberg 2012

the curls; thus water has eddying motions, one part of which is due to the principle current, the other to the random and reverse motion."
    L. da Vinci 1510

2.1.2   The Eddy-Viscosity Hypothesis/Boussinesq Assumption

The eddy viscosity hypothesis/ Boussinesq assumption is that:

> The small eddies act to drain energy from the large eddies.

This description is quite old in fluid mechanics. For example, Venturi in 1797 wrote[1] about retardation of a flow caused by eddies of different velocities interacting and Boussinesq gave an early and compelling theoretical argument in favor of it. Paraphrasing Prandtl's 1931 description, the eddy viscosity hypothesis is that a

> fluid flowing turbulently behaves, in the mean, like a fluid of greatly increased and highly variable viscosity.

The *Eddy Viscosity Hypothesis* thus seeks to model the energy lost to the resolved scales when two eddies interact and break down into ones smaller than the cutoff length scale. This process of breakdown of eddies into smaller and smaller ones is often called the Richardson energy cascade since it was described by L.F. Richardson in his famous poem. There is considerable evidence in real fluids for those two physical ideas. Their mathematical derivation from the Navier–Stokes equations is still, however, an open question. We shall use these two ideas to present the family of models we are considering.

Averaging typically is a smoothing operator, eliminates any local oscillations smaller than $O(\delta)$, and commutes with differentiation. A smoothing averaging process is often depicted (below in Fig. 2.1) as decomposing a wiggly function, u, into its mean (bold curve below), $\overline{u}$, and its fluctuation about the mean (the dashed curve fluctuating about the x-axis), $u' := u - \overline{u}$.

Exact spacial averaging suppresses small spacial scales. It is widely believed that spacial averaging (through the space and time coupling in the equations of motion) also suppresses correspondingly small time scales. This belief certainly corresponds to our everyday observation of fluids in motion: spacial averaging (done by looking at a river from a greater distance) seems to make it flow more slowly. This was well described by Wordsworth.

---

[1]In 1838 Saint-Venet called this extra retardation *extraordinary friction* and others had called it *loss of live force* where *live force* was the term used for kinetic energy.

Fig. 2.1 A curve, its mean (*heavy line*) and fluctuation (*dashed*)

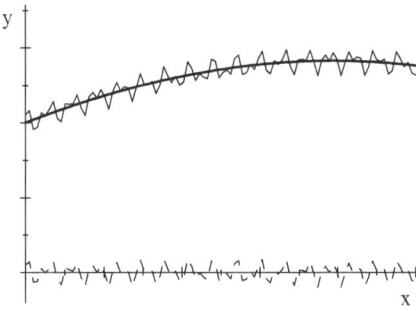

"Yon foaming flood seems motionless as ice,
Its dizzy turbulence eludes the eye,
Frozen by distance."
W. Wordsworth, 1770–1850,
in: Address to Kilchurn Castle.

2.2 Local Spacial Averages

The basic philosophy in large eddy modeling is based on the physical ideas of sensitivity and scale separation stated above, and there is considerable evidence in real fluids for these ideas. Their mathematical derivation from the Navier–Stokes equations is still, however, an open question. We shall accept these two ideas to present the family of approximate deconvolution models we are considering. Let $g(x)$ be a smooth function with, $0 \leq g \leq 1$, $g(0) = 1$ and $\int_{\mathbf{R}^d} g \, dx = 1$. Filtering requires a filter kernel. A filter kernel or mollifier $g_\delta(x)$ is defined (as above) by

$$g_\delta(x) := \delta^{-d} \, g(x/\delta), \text{ where } d = \text{dimension}(\Omega) = 2 \text{ or } 3 \ .$$

The function $g(x)$ is assumed to have compact support or, at least, exponential decay. A typical example is a Gaussian:

$$g(x) = (6/\pi)^{d/2} \, \exp(-6 \, |x|^2),$$

depicted in Fig. 2.2, and next for $\delta = 1$ (bold curve) and rescaled by $\delta = \frac{1}{3}$ (thin curve) in Fig. 2.3.

The local spacial filter is defined by convolution with g_δ. Thus, given $u(x)$ define the local average $\bar{u}$ and fluctuation u' by

$$\bar{u}(x) = (g_\delta * u)(x) := \int_{\mathbf{R}^d} g_\delta(x - y)u(y) \, dy, \text{ and } u' = u - \bar{u},$$

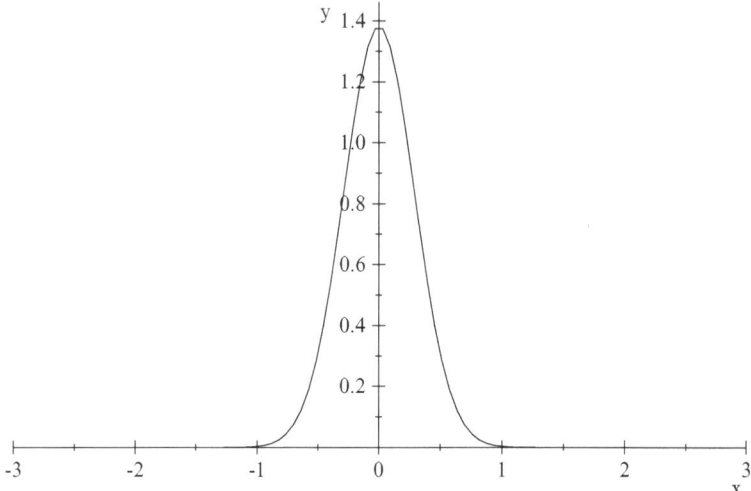

Fig. 2.2 The Gaussian in one dimension

Fig. 2.3 A gaussian filter
(*heavy*) and rescaled (*thin*)

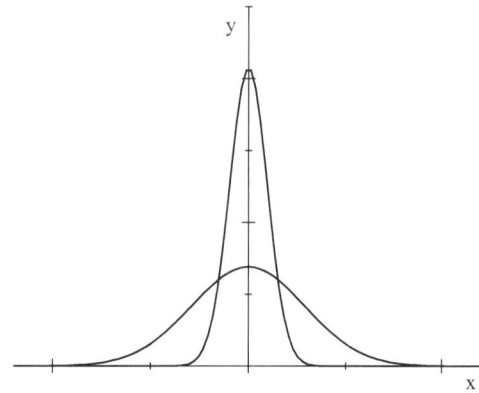

where u is extended by zero off Ω. This convolution is a smoothing operator. It eliminates any local oscillations smaller than $O(\delta)$. Convolution has many other important mathematical properties. For example, in the absence of boundaries it commutes with differentiation so

$$g_\delta * \left(\frac{\partial}{\partial x_j} u\right) = \frac{\partial}{\partial x_j}(g_\delta * u).$$

Ignoring boundaries for the moment and using these nice properties of convolution, apply this filter to the Navier–Stokes equations; i.e., take $g_\delta * NSE$ ($u = g_\delta * f$). This yields the exact SFNSE = Space Filtered Navier

Stokes Equations

$$\bar{u}_t + \nabla \cdot (\overline{uu}) + \nabla \bar{p} - Re^{-1}\Delta \bar{u} = \bar{f}, \; \nabla \cdot \bar{u} = 0. \qquad (2.1)$$

Traditionally, and for good reasons, the nonlinear term is rewritten as

$$\overline{uu} = \bar{u}\bar{u} - (\bar{u}\bar{u} - \overline{uu}),$$

so that the SFNSE becomes an equation for $\bar{u}$ which looks like the NSE plus one additional stress term:

$$\bar{u}_t + \nabla \cdot (\bar{u}\,\bar{u}) + \nabla \bar{p} - Re^{-1}\Delta \bar{u} - \nabla \cdot \mathbf{R}(u) = \bar{f}, \qquad (2.2)$$

$$\nabla \cdot \bar{u} = 0, \qquad (2.3)$$

$$\text{where} \quad \mathbf{R}(u) := \bar{u}\,\bar{u} - \overline{u\,u}. \qquad (2.4)$$

The tensor $\mathbf{R}(u)$, called the sub-filter scale stress tensor, represents the stress the unresolved scales exerts upon the resolved scales. Since both $\overline{uu}$ and $\mathbf{R}$ in either form of the SFNSE are functions of u and not $\bar{u}$, the exact SFNSE system is not closed. Various closure models are used to obtain a system that can be discretized and solved.

Many other averages are used in LES due to the wide variety of algorithms and applications in which turbulence is prominent. Another disadvantage of the Gaussian is that it must be truncated as otherwise the Gaussian filter requires an integral over all of $\mathbb{R}^3$. More local averages are often convenient. One example is the top hat filter.

2.2.1 Top Hat Filter

The top hat filter is the un-weighted average defined over a neighborhood of a given point:

$$B_\delta(x) := \{y : |x - y| < \delta\},$$

$$\bar{u}(x) := \frac{1}{vol(B_\delta(x))} \int_{B_\delta(x)} u(y)dy.$$

Thus in $3d$ this means

$$\bar{u}(x) := \frac{1}{\frac{4}{3}\pi\delta^3} \int_{|x-y|<\delta} u(y)dy.$$

This can be written as a convolution by choosing $g_\delta(x) := \delta^{-3}g(x/\delta)$ where

$$g(x) = 1 \text{ , if } |x| < \frac{3}{4\pi},$$

$$g(x) = 0 \text{ , if } |x| \geq \frac{3}{4\pi}.$$

2.2.2 Discrete Filters

In finite difference approximations, ultimately one must filter discrete velocities defined on a finite difference mesh. On a uniform mesh in $2d$ (chosen to make figures easier) using the standard finite difference compass notation the analog of the top hat filter is

$$\overline{u}(P) := \frac{u(N) + u(S) + u(E) + u(W) + u(P)}{5}.$$

2.2.3 Weighted Discrete Filters

The above shares many of the drawbacks of the top hat filter. Weighted local averages are more commonly used. As an example of how to develop weighted averages, suppose that at each mesh point P, $\overline{u}$ satisfies

$$-\delta^2 \frac{\overline{u}(N) + \overline{u}(S) - 4\overline{u}(P) + \overline{u}(E) + \overline{u}(W)}{h^2} + \overline{u}(P) = u(P). \tag{2.5}$$

This is a coupled linear system for the filtered velocity. It's condition number is $O((\frac{\delta}{h})^2 + 1)$ so it is quite reasonably to do one or two relaxation steps to define a new filtered velocity. For example, one step of Jacobi gives

Guess $\overline{u}^{old}$

$$\overline{u}^{new}(P) := \frac{(\frac{\delta}{h})^2\overline{u}^{old}(N) + (\frac{\delta}{h})^2\overline{u}^{old}(S) + (\frac{\delta}{h})^2\overline{u}^{old}(E) + (\frac{\delta}{h})^2\overline{u}^{old}(W) + u(P)}{1 + 4(\frac{\delta}{h})^2}$$

If $\delta = h$ and only one step is performed we return to the discrete top hat filter.

$$\overline{u}(P) := \frac{u(N) + u(S) + u(E) + u(W) + u(P)}{5}$$

2.2.4   Other Filters

Other filters can be obtained by:

- Different choices of δ, e.g., $\delta = 3h$.
- Other relaxation methods.
- More steps of the relaxation method chosen.

We refer the interested reader to Vasilyev et al. [VLM98, HV03] and references therein for more in depth literature on discrete filters.

2.2.5   Weighted Compact Discrete Filter from [SAK01a]

There has developed a considerable experience with weighted discrete filters inspired by the needs of difference methods. On structured meshes, filters can be derived in 1d and extended by taking tensor products of 1d filters. Generally, the higher order the filter, the more points involved in the averaging operator and thus the greater the bandwidth on the linear system that must be solved. Compact filters are a clever idea of Stolz, Adams and Kleiser [SAK01a]; they attain higher order but only require tridiagonal solves (on structured meshes). The following 1d weighted discrete filter from [SAK01a] has second order accuracy and has proven its value in LES. Given values u_i of the variable u at equi-spaced mesh points x_i, a weighting parameter α is chosen in the range $-1/2 \leq \alpha \leq +1/2$. Then, filtered values are calculated by solving the tridiagonal system

$$\alpha \overline{u}_{i-1} + \overline{u}_i + \alpha \overline{u}_{i+1} = \left(\frac{1}{2} + \alpha \right) \left(u_i + \frac{u_{i-1} + u_{i+1}}{2} \right).$$

One measure of success is the number of extensions and generalizations. Visbal and Gaitonde [VG02] gave the following 2Mth order compact extension of Stolz, Adams and Kleiser's compact filter idea:

$$\alpha \overline{u}_{i-1} + \overline{u}_i + \alpha \overline{u}_{i+1} = \sum_{n=0}^{M} \left(\frac{a_n}{2} \right) (u_{i-n} + u_{i+n}).$$

Note that this filter still only requires a tridiagonal system solve. The values of the weighting coefficients on the RHS are given in Table 2.1.

Table 2.1 Coefficients of 2Mth order compact filters

·	2nd_order	4th_order	6th_order	8th_order
a_0	$\frac{1}{2} + \alpha$	$\frac{5}{8} + \frac{3\alpha}{4}$	$\frac{11}{16} + \frac{5\alpha}{8}$	$\frac{93}{128} + \frac{70\alpha}{128}$
a_1	$\frac{1}{2} + \alpha$	$\frac{1}{2} + \alpha$	$\frac{15}{32} + \frac{17\alpha}{16}$	$\frac{7}{16} + \frac{18\alpha}{16}$
a_2	0	$-\frac{1}{8} + \frac{\alpha}{4}$	$-\frac{3}{16} + \frac{3\alpha}{4}$	$-\frac{7}{32} + \frac{14\alpha}{32}$
a_3	0	0	$\frac{1}{32} - \frac{\alpha}{16}$	$\frac{1}{16} - \frac{\alpha}{8}$
a_4	0	0	0	$-\frac{1}{128} + \frac{\alpha}{64}$

2.2.6 Differential Filters

Differential filters are well-established in large eddy simulation, starting
with work of Germano [Ger86] and continuing [GL00, S01], and have many
connections to regularization processes such as the Yoshida regularization of
semigroups and was used in the very interesting work of Foias, Holm, Titi
[FHT01] (and others) on the NS-α model. Filtering with a differential filter is
also one of the simplest elliptic-elliptic singular perturbation problems. If the
boundary conditions are periodic or if the unfiltered and filtered velocities
are equal on the boundary (e.g., both vanish) then the filtered velocity
should have no significant boundary layers. This is reflected in the analysis of
differential filters: it is simple and gives useful and strong results quite easily.

In the case of periodic boundary conditions, given 2π- periodic divergence-
free field w with zero mean, its spacial average over $O(\delta)$ length-scales,
denoted $\overline{w}$, is the unique 2π- periodic solution of the Stokes problem

$$-\delta^2 \triangle \overline{w} + \overline{w} + \nabla \pi = w \quad \text{in } \mathbb{R}^3, \quad \nabla \cdot \overline{w} = 0, \quad \int_\Omega \overline{w} = 0. \quad (2.6)$$

It can be shown that π is a constant in the equation above and therefore the
pressure term disappears. Thus in the periodic case we have: given $w \in L^2(\Omega)$,
its average, denoted $\overline{w}$, is the solution with $\overline{w} \in L^2(\Omega)$ and $\nabla \overline{w} \in L^2(\Omega)$ of

$$-\delta^2 \triangle \overline{w} + \overline{w} = w.$$

The definition of the differential filter must be modified for the problem
with no-slip boundary conditions. Briefly, the above definition preserves
incompressibility for periodic boundary conditions and for the Cauchy
problem but not for the problem with no-slip boundary conditions. The
correct modification is to solve a shifted Stokes problem. Given $\phi \in L^2(\Omega)$,
with $\nabla \phi \in L^2(\Omega)$ its average, denoted $\overline{\phi}$, is the solution with $\overline{\phi} \in L^2(\Omega)$, and
also $\nabla \overline{\phi} \in L^2(\Omega)$ of the following problem: find $\overline{\phi}, \lambda$ satisfying

$$-\delta^2 \{\triangle \overline{\phi} + \nabla \lambda\} + \overline{\phi} = \phi, \text{ in } \Omega,$$

$$\nabla \cdot \overline{\phi} = \nabla \cdot \phi \text{ , in } \Omega,$$

$$\overline{\phi} = \phi \text{ , on } \partial\Omega.$$

Generally, differential filtering shares many features of an averaging process; it defines a positive semi-definite operator which is smoothing, attenuates high frequencies and converges to the identity operator as $\delta \to 0$, i.e., $\overline{\phi} \to \phi$ as $\delta \to 0$. Much of the mathematical foundation of filtering is phrased in terms of the transfer function or symbol of the filtering operator under Fourier transform or Fourier series.

In the periodic case, let $\overline{w}$ be the unique periodic solution to

$$-\delta^2 \triangle \overline{w} + \overline{w} = w. \tag{2.7}$$

Writing

$$w(\mathbf{x}, t) = \sum_{\mathbf{k}} \widehat{w}(\mathbf{k}, t) e^{-i\mathbf{k}.\mathbf{x}}, \text{ we have}$$

$$\overline{w}(\mathbf{x}, t) = \sum_{\mathbf{k}} \frac{\widehat{w}(\mathbf{k}, t)}{1 + \delta^2 |\mathbf{k}|^2} e^{-i\mathbf{k}.\mathbf{x}}.$$

Then writing $\overline{w} = G(w)$, we see that, the transfer function of G, denoted by $\widehat{G}$ is the function

$$\widehat{G}(\mathbf{k}) = \frac{1}{1 + \delta^2 |\mathbf{k}|^2}.$$

Moreover, the transfer function depends only on the modulus of the wave vector $\mathbf{k}$. Therefore, with $k = |\mathbf{k}|$, we shall write in the following $\widehat{G}(k)$ instead of $\widehat{G}(\mathbf{k})$. Normally this is represented graphically by the obvious re-scaling $\delta k \to k$, so $\widehat{G}(k) = (1 + k^2)^{-1}$ (Fig. 2.4).

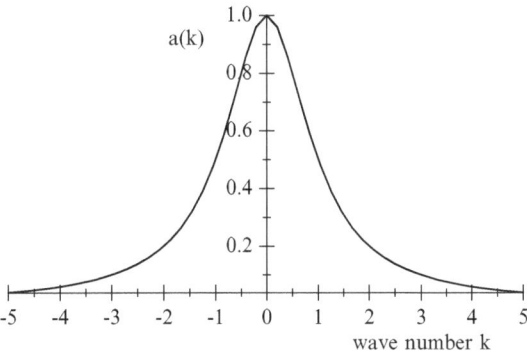

Fig. 2.4 Transfer function of the differential filter

The smoothing property of the differential filter is reflected in the decay at infinity of its transfer function.

It follows readily that the averaging process is stable and smoothing in the sense below and the filtered function is a good approximation of the true function, if it is smooth.

Lemma 8 (Stability and smoothing of differential filters). *Suppose periodic boundary conditions hold. For any $\phi \in L^2(\Omega)$,*

$$||\overline{\phi}||_{L^2(\Omega)} \leq ||\phi||_{L^2(\Omega)}, \tag{2.8}$$

$$\sqrt{2}\delta||\nabla\overline{\phi}||_{L^2(\Omega)} \leq ||\phi||_{L^2(\Omega)}, \; and \tag{2.9}$$

$$\frac{1}{2}\delta^2||\triangle\overline{\phi}||_{L^2(\Omega)} \leq ||\phi||_{L^2(\Omega)}. \tag{2.10}$$

Proof. Multiply (1.5) by $\overline{\phi}$, integrate over Ω and integrate by parts. This gives

$$\int_{\Omega} \delta^2|\nabla\overline{\phi}|^2 + |\overline{\phi}|^2 \; dx \leq \int_{\Omega} \phi\overline{\phi} \; dx.$$

Use the Cauchy–Schwarz inequality on the right hand side gives the first two. For the third, note that the definition of the differential filter implies $-\delta^2\triangle\overline{\phi} = \phi - \overline{\phi}$. Thus,

$$\delta^2||\triangle\overline{\phi}||_{L^2(\Omega)} = ||\phi - \overline{\phi}||_{L^2(\Omega)} \leq 2||\phi||_{L^2(\Omega)}.$$

$\square$

Lemma 9 (Accuracy of approximation by filtering). *Suppose periodic boundary conditions hold. For any $\phi \in L^2(\Omega)$,*

$$||\phi - \overline{\phi}||_{L^2(\Omega)} \leq \frac{1}{\sqrt{2}}\delta||\nabla\phi||_{L^2(\Omega)}, \tag{2.11}$$

$$||\nabla(\phi - \overline{\phi})||_{L^2(\Omega)} \leq \frac{1}{\sqrt{2}}\delta||\triangle\phi||_{L^2(\Omega)}, \; and \tag{2.12}$$

$$||\phi - \overline{\phi}||_{L^2(\Omega)} \leq \delta^2||\triangle\phi||_{L^2(\Omega)}. \tag{2.13}$$

Further, for any derivative $\partial^{|\beta|}\phi/\partial x^\beta$, with multi-index $|\beta| \geq 0$

$$\left|\left|\frac{\partial^{|\beta|}}{\partial x^\beta}(\phi - \overline{\phi})\right|\right| = \delta^2\left|\left|\triangle\frac{\partial^{|\beta|}}{\partial x^\beta}\overline{\phi}\right|\right| \leq \delta^2\left|\left|\triangle\frac{\partial^{|\beta|}}{\partial x^\beta}\phi\right|\right|, \; and$$

$$\left|\left|\frac{\partial^{|\beta|}}{\partial x^\beta}(\phi - \overline{\phi})\right|\right| \leq \frac{\delta}{\sqrt{2}}\left|\left|\nabla\frac{\partial^{|\beta|}}{\partial x^\beta}\phi\right|\right|.$$

Proof. The averaging error by $\Phi = (\phi - \overline{\phi})$ satisfies the equation

$$-\delta^2 \triangle \Phi + \Phi = -\delta^2 \triangle \phi.$$

The above error bounds for Φ follow in much the same ways as the above stability bounds.   □

The first derivative estimate follows from the definition of averaging equation, stability of averaging and the fact that derivatives commute under periodic boundary conditions. For the second, note that the error equation satisfies

$$-\delta^2 \triangle (\phi - \overline{\phi}) + (\phi - \overline{\phi}) = -\delta^2 \triangle \phi.$$

Differentiating through this equation shows that the derivatives of the error also satisfy the same equation (with the derivative of ϕ on the RHS instead of ϕ). Multiplying by $(\phi - \overline{\phi})$, integrating over Ω, integrating by parts and using the Cauchy - Schwarz inequality gives

$$\delta^2 ||\nabla(\phi - \overline{\phi})||^2 + ||(\phi - \overline{\phi})||^2 \leq \delta^2 ||\nabla\phi|| ||\nabla(\phi - \overline{\phi})||$$

$$\leq \delta^2 ||\nabla(\phi - \overline{\phi})||^2 + \frac{\delta^2}{4} ||\nabla\phi||^2,$$

and the result follows for the error $(\phi - \overline{\phi})$. Since the derivatives of the error satisfy the same equations as the error, the same proof works for the derivatives of the error as well.

2.2.7 *Scale Space: What Is the Right Averaging?*

Large eddy simulation is concerned with approximation of local, spacial averages in under-resolved simulations of mixes of laminar, transitional and turbulent flows. Traditionally, velocity averages $\overline{u}$ in LES are defined by convolution with a given filter

$$\overline{u}(x, t; \delta) := (g_\delta \star u)(x, t), \delta = \text{ filter radius, typically } O(h).$$

On the surface, there are many choices of filter kernel $g(\cdot)$. However, if the averages are viewed as containing information of physical meaning, there are already results on acceptable filters beginning with the famous paper of Koenderink [Koe84], see also Sagaut [S01]. Scale space analysis begins with some basic postulates that any physically reasonable filter should satisfy. In [Koe84] (see also [Lin94] for interesting developments and applications) the following is proven.

Theorem 10 (Koenderink [Koe84]). *The only filter $\overline{u}(x,t) := g_\alpha \star u(x,t)$ satisfying the following four conditions is the Gaussian filter:*

linearity:

$$\overline{\alpha u + \beta v} = \alpha \overline{u} + \beta \overline{v}$$

spacial invariance:

$$\overline{u(x - a)} = \overline{u}(x - a)$$

isotropy, for all rotations R :

$$\overline{u}(x) = R^* \overline{u(Rx)}$$

scale invariance (no preferred size):

as $\overline{u}(x,t) := g_\alpha \star u(x,t)$, $\overline{\overline{u}} := g_\alpha \star g_\alpha \star u$, then $\overline{\overline{u}} = g_{\sqrt{2}\alpha} \star u$.

Thus, there is really only one mathematical correct filter: the Gaussian. Convolution with the Gaussian is expensive so that in spite of this strong uniqueness result usually other filters are used. Koenderink's result indicates that these other filters must be assessed as approximations to the Gaussian.

Differential filters were proposed for large eddy simulation by Germano [Ger86]. The close connection of the above differential filter to the Gaussian filter can be seen various ways, [BIL06]. For example, the Gaussian is the heat kernel. Thus, one way to compute the exact Gaussian filtered velocity $\overline{u}(x,t) = g_\delta \star u(x,t)$ is to solve the following evolution equation:

$$v_s(x,s) = \triangle v(x,s) \text{ for } s > 0, \text{ and } v(x,0) = u(x),$$

then set $\overline{u}(x,t) := v(x,s)|_{s=\delta^2}$. Since the averaging radius δ is small, δ^2 is smaller still and we can reasonably approximate $v(x,\delta^2)$ by one step of backward Euler, leading back to the differential filter.

In comparison with the Gaussian, the differential filter is only approximately scale invariant. Indeed, if we compute $\overline{\overline{u}}$ it fails the semigroup property by $O(\delta^4)$,

$$\overline{\overline{u}} = (-\delta^2\triangle + 1)^{-1}(-\delta^2\triangle + 1)^{-1}u = (-(\sqrt{2}\delta)^2\triangle + 1)^{-1}u + O(\delta^4).$$

2.3 The SFNSE

In developing turbulence models, the nonlinear term $u \cdot \nabla u$ is commonly rewritten in an equivalent way. Define the tensor $u \otimes u$ (which in fluids circles is usually written uu) by

$$(uu)_{ij} = (u \otimes u)_{ij} := u_i u_j.$$

Then, using $\nabla \cdot u = 0$ it follows that

$$u \cdot \nabla u = \nabla \cdot (uu).$$

Let $g(x)$ be a smooth function with, $0 \le g \le 1$, $g(0) = 1$ and $\int_{\mathbb{R}^d} g \, dx = 1$. The mollifier $g_\delta(x)$ is defined (as usual) by

$$g_\delta(x) := \delta^{-d} \, g(x/\delta).$$

The local spacial filter is defined by convolution with g_δ. Thus, given $u(x)$ define:

$$\bar{u}(x) = (g_\delta * u)(x) := \int_{\mathbb{R}^d} g_\delta(x - y)u(y)dy, \text{ and } u' = u - \bar{u}.$$

Ignoring boundaries for the moment and using these nice properties of convolution, apply this filter to the Navier–Stokes equations (i.e., take $g_\delta * NSE(u) = g_\delta * f$). This gives the *SFNSE=Space-Filtered-NSE*

$$\bar{u}_t + \nabla \cdot (\overline{u\,u}) - \nu\triangle\bar{u} + \nabla\bar{p} = \bar{f}, \text{ and } \nabla \cdot \bar{u} = 0, \tag{2.14}$$

or, after rearranging into the most commonly seen form,

$$\bar{u}_t + \nabla \cdot (\bar{u}\,\bar{u}) + \nabla\bar{p} - \nu\triangle\bar{u} + \nabla \cdot \mathbf{R}(u, u) = \bar{f}, \ \nabla \cdot \bar{u} = 0, \tag{2.15}$$

where $\mathbf{R}(u, u)$ is the tensor representing the stress the unresolved scales exerts upon the resolved scales:

$$\mathbf{R}(u, u) := \overline{u\,u} - \bar{u}\,\bar{u}.$$

The tensor $\mathbf{R}(u, u)$ is often called the *sub-filter scale stress tensor* and it is sometimes called the Reynolds stress tensor. (There is a terminological debate about whether the latter is correct or a misnomer.) Since $\mathbf{R}$ is a function of u and not only of $\bar{u}$, this system is not closed. Two ways (of many) to describe the closure problem are to replace the tensor (a) uu in the SFNSE with a tensor $\mathbf{S}(\bar{u}, \bar{u})$ that depends only on $\bar{u}$ not u and (b) to do the same instead with $\mathbf{R}(u, u)$.

If we start by the approach (a), calling w, q the approximations to u, p that result from replacing the tensor uu in the SFNSE with a tensor $S(\bar{u}, \bar{u})$, this leads to the following large eddy simulation model for predicting the averages $\bar{u}, \bar{p}$

$$w_t + \nabla \cdot (\overline{\mathbf{S}(w, \ w)}) - \nu\triangle w + \nabla q = \bar{f}, \text{ and } \nabla \cdot w = 0.$$

The true filtered momentum equation can be rewritten as

$$\overline{u}_t + \nabla \cdot (\overline{\mathbf{S}(\overline{u},\ \overline{u})}) - \nu \triangle \overline{u} + \nabla \overline{p} = \overline{f} + \overline{\nabla \cdot \tau}.$$

Motivated by this form of the SFNSE, the consistency error tensor is $\tau :=$ $S(\overline{u},\ \overline{u}) - u\ u$. Comparing the last two equations, the deviation of the true flow averages $\overline{u}$ from the model's solution w is driven by the consistency error tensor τ.

Definition 11 (Modeling and numerical errors). The modeling error is the deviation of the model's solution from the true flow averages

$$e_{modeling} := \overline{u} - w.$$

Let w^h denote a computed approximation of the large eddy simulation model. The numerical error is the deviation of the computed solution from the large eddy simulation model's true solution,

$$e_{numerical} := w - w^h.$$

Subtracting the last two equations (SFNSE-LESmodel) gives the equation for the modeling error, $e(x,t) = \overline{u} - w$: $e(x,0) = 0, \nabla \cdot e = 0$ and

$$(\overline{u} - w)_t + \nabla \cdot (\overline{\mathbf{S}(\overline{u},\ \overline{u})} - \overline{S(w,\ w)}) - \nu \triangle (\overline{u} - w) + \nabla (\overline{p} - q) = \overline{\nabla \cdot \tau}, \tag{2.16}$$

which is driven only by the model's consistency error τ. Thus, having a small modeling error depends on (a) having a small model consistency error, and (b) the model being stable and stable to perturbations.

2.4   Eddy Viscosity Models

"Turbulent fluid behaves like a fluid of greatly increased viscosity, with the difference, however, that the increase of the viscosity varies considerably from place to place." -Prandtl,1931.

"It has been reported.... that when a Newtonian fluid flows down a straight pipe of non-circular cross-section, under conditions for which the fluid has become fully turbulent, the mean flow is no longer rectilinear, but a secondary flow exists in the cross-sectional planes of a type similar to that can be calculated for some non-Newtonian fluids. This fact suggests that the turbulent Newtonian liquid may, for certain purposes, be regarded as a non-Newtonian fluid."
Rivlin

The second approach to closure is to replace the tensor $\mathbf{R}(u, u)$ in the rearranged SFNSE with a tensor that depends only on $\overline{u}$ not u. The simplest

and most commonly used closure assumption is the eddy viscosity hypotheses
that the turbulent stress is a linear function of the large scales' deformation
tensor $\nabla^s \bar{u}$. This model is:

$$\nabla \cdot \mathbf{R}(u, u) \sim -\nabla \cdot (2\nu_T \, \nabla^s \bar{u}) + \text{terms incorporated into the pressure,}$$

$$\nu_T := \text{ turbulent viscosity coefficient.}$$

This is called the Boussinesq assumption or eddy viscosity hypothesis. It is
a modeling assumption that is easily checked to be false in its details but
is hoped to capture some of the important effects of the unresolved scales
on the mean flow. To be most precise, we must split $\mathbf{R}$ into two parts. One,
the average of the stresses in the $x - y - z$ directions, is incorporated into
the turbulent pressure and the other, the trace free part of $\mathbf{R}$, is modeled by
turbulent diffusion

$$\mathbf{R} = \left(\mathbf{R} - \frac{1}{3} trace(\mathbf{R})\mathbf{I} \right) + \frac{1}{3} trace(\mathbf{R})\mathbf{I}.$$

Thus the mean turbulent pressure $\bar{p}$ in the SFNSE should really be defined
to be

$$\bar{p} \Leftarrow \bar{p} + \frac{1}{3} trace(\mathbf{R}).$$

Eddy viscosity thus most properly and precisely models only the zero trace
part of $\mathbf{R}$

$$\nabla \cdot \left(\mathbf{R} - \frac{1}{3} trace(\mathbf{R})I \right) \sim -\nabla \cdot (\nu_T \nabla^s \bar{u})$$

$$\nu_T := \text{turbulent viscosity coefficient.}$$

Thus, when the eddy viscosity hypothesis is used, the solution sought is no
longer the exact $(\bar{u}, \bar{p})$ but rather an approximation to $(\bar{u}, \bar{p} + \frac{1}{3} trace(\mathbf{R}))$.
Calling the approximation (w, q), the resulting model becomes:

$$w_t + \nabla \cdot (ww) + \nabla q - \nabla \cdot (2[\nu + \nu_T]\nabla^s \bar{u}) = \bar{f}, \text{ in } \Omega,$$

$$\nabla \cdot w = 0, \text{ in } \Omega. \qquad (2.17)$$

There are many models for determining the turbulent viscosity coefficient

$$\nu_T = \nu_T(\delta, \, \bar{u}).$$

The most basic requirement is the eddy viscosity be dimensionally consistent
with viscosity. Second, it seems plausible that the amount of turbulent mixing

should depend on the local cutoff length scale and on the kinetic energy of the turbulent fluctuations doing the mixing. The dimensionally consistent formula determined is called the *Kolmogorov–Prandtl relation*:

$$\nu_T = \nu_T(l,\ k') = Cl\sqrt{k'}$$

$$l = \text{mixing length},$$

$$k' = \text{turbulent kinetic energy (TKE)}.$$

For example, in LES one can simply take

$$l = \delta\ ,\ (\text{averaging radius}),$$

$$k' \simeq \frac{1}{2}|u - \overline{u}|^2,\ (\text{kinetic energy in fluctuations}),$$

$$\nu_T = C\delta|u - \overline{u}|.$$

Unfortunately, u is unknown so further approximations of k' must be made. Thus, closure (in this approach) reduces to predicting the TKE in terms of resolved quantities.

2.4.1   A First Choice of ν_T

"With four parameters I can fit an elephant, and with five I can make him wiggle his trunk."

J. von Neumann, quoted by Freeman Dyson in p. 297: *A meeting with Enrico Fermi*, Nature 427 (22 January 2004).

There are many models for determining the turbulent viscosity coefficient $\nu_T = \nu_T(\delta,\ \overline{u})$. The simplest choice of ν_T is a constant. One goal of an LES model is to make the eddies smaller than $O(\delta)$ vanish. In other words to make the cutoff length scale for the LES model comparable to some feasible meshwidth. This is done by increasing the energy dissipation in the model. This is not a good choice because it just reduces a turbulent flow to a laminar one. However, it introduces some interesting ideas so we shall consider it first. With this in mind, the eddy viscosity model describes a flow with effective Reynolds number

$$\text{Effective Reynolds Number} := (Re^{-1} + \nu_T)^{-1} \sim \nu_T^{-1}.$$

The K41 theory of Kolmogorov thus predicts the length scale of the smallest persistent eddy in solutions of the model to be

$$\eta_{LES} \doteq O([Re^{-1} + \nu_T]^{3/4}).$$

We wish $\eta_{LES} = O(\delta)$, so, in the interesting case in which $Re^{-1} << O(\delta)$, this means $\nu_T = O(\delta^{4/3})$. This yields the global choice

$$\nu_T = C\delta^{4/3}$$

where $C > 0$ is an $O(1)$ fitting constant. The problem with this constant choice of ν_T is the model's action on the smooth parts of the flow - it is just too diffusive! The flow that results is laminar. This observation motivates efforts to find better models which have similar effects on the small eddies but do not distort the large eddies too much.

2.5 The Smagorinsky Model

"The road to full knowledge of the variations of viscosity appears to lie in the study of diffusion of eddies."
L.F. Richardson, 1922, p. 79 in [R22].

"The only encouraging prospect is that current progress in understanding turbulence will restrict the freedom of such modeling and guide these efforts toward a more reliable discipline."
H. W. Liepman, 1979, p. 221 in: *American Scientist*, vol. 62.

In 1963 Smagorinsky proposed the choice for ν_T which is still the most popular today[2]:

$$\nu_T = (C_S\ \delta)^2 |\nabla^s w|_F. \tag{2.18}$$

This model was advanced independently by Ladyzhenskaya for other reasons. The same regularization of the compressible Navier–Stokes equations had been used in the 1950's by Richtmeyer and von Neumann for computations of compressible flows with shocks. It is also the most mathematically appealing choice for ν_T.

Let us now consider a choice like the above:

$$\nu_T = C_s\ \delta^r\ |\nabla^s w|_F^s,$$

$$|\nabla^s w|_F := \sqrt{\sum_{i=1,2,3} \sum_{j=1,2,3} \left(\frac{w_{i,j} + w_{j,i}}{2}\right)^2}$$

---

[2] Recall that the Frobenius norm of a tensor is given by
$|\nabla w|_F^2 = \sum_{i,j=1,2,3} \left(\frac{\partial w_j}{\partial x_i}\right)^2.$

where r and s must be determined. Following the previous reasoning, the *effective local Reynolds number* is:

$$Re_{Effective} = (Re^{-1} + C_s \, \delta^r |\nabla^s w(x)|^s)^{-1} \sim (C_s \, \delta^r |\nabla^s w(x)|^s)^{-1}.$$

If we assume the K41 theory can be applied *locally* as well as globally after time averaging(and there is little evidence either for or against this assumption, but to make progress something must be assumed) this gives the *local* length scale of:

$$\eta_{LES}(x) \doteq O([Re^{-1} + C_s \, \delta^r |\nabla^s w(x)|_F^s]^{3/4}) \doteq O(\delta^{(3/4)r} |\nabla^s \mathbf{w}|^{3/4 \, s}(x)),$$

in the most interesting case when Re^{-1} is negligible.

If x is chosen to lie in an eddy of size $\leq O(\delta)$ then $|\nabla^s w(x)| \geq O(\delta^{-1})$. Thus, if we want

$$\min_x \eta_{LES}(x) = \delta,$$

this gives:

$$\delta \doteq \delta^{3/4r} \, \delta^{-3/4s}, \text{ or } r = \frac{4}{3} + s.$$

This choice of r and s gives an attractive family of LES models:

$$\nu_T(\delta, w) := C_s \, \delta^{4/3+s} |\nabla^s w|^s, \text{ where } s > 0, \tag{2.19}$$

yielding the model

$$w_t + \nabla \cdot (w \, w) + \nabla q, -\nabla \cdot [2(\, Re^{-1} + C_s \, \delta^{4/3+s} |\nabla^s w|^s) \nabla^s w] = \bar{f},$$
$$\nabla \cdot w = 0. \tag{2.20}$$

For example, for $s = 1$, this gives

$$\nu_T(\delta, w) := C_s \, \delta^{7/3} |\nabla^s w|, \tag{2.21}$$

which is a different value than used in the Smagorinsky model (considered next). For the large eddies (when $|\nabla^s w| = O(1)$), we can make the influence of the turbulent viscosity smaller by picking s larger. We expect that the eddies smaller than $O(\delta)$ are also rapidly attenuated. Thus, (2.20) is an important improvement in LES models over constant ν_T. However, the constant C_s in this alternate scaling is no longer non dimensional. It must therefore either depend on δ (which obviates the above calculation) or on L (whose choice is somewhat arbitrary).

The Smagorinsky model corresponds to the special choice $r = 2, s = 1$. This choice makes C_s dimension free as a first step to a universal model. The Smagorinsky model is given by

$$w_t + \nabla \cdot (w\ w) + \nabla q, -\nabla \cdot [2(\ Re^{-1} + (C_S\ \delta)^2 |\nabla^s w|_F) \nabla^s] = \bar{f},$$

$$\nabla \cdot w = 0, \text{ in } \Omega.$$

$$(2.22)$$

Variations on the Smagorinsky model have proven to be the workhorse in large eddy simulations of industrial flows. Variants are needed because the model as presented, while far better than constant eddy viscosity, is still far too dissipative. For example, here is a simple test of the Smagorinsky model for $2d$ flow over a step (far from the case of turbulence). At the Reynolds number of this test, around 600, eddies are periodically shed from the step and roll down the channel, Fig. 2.5; the flow does not approach a steady state.

However, if this simulation is performed on a coarser mesh, it makes sense to try the Smagorinsky model to see if it gives the correct large structures. Here is the result which is clearly over-diffused: eddies are not shed and the

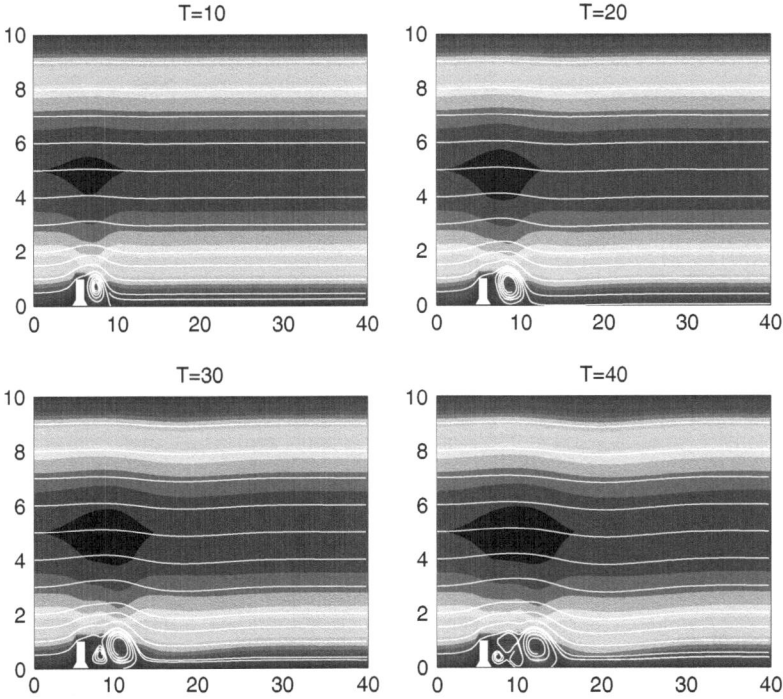

Fig. 2.5 Eddies are shed and roll down channel

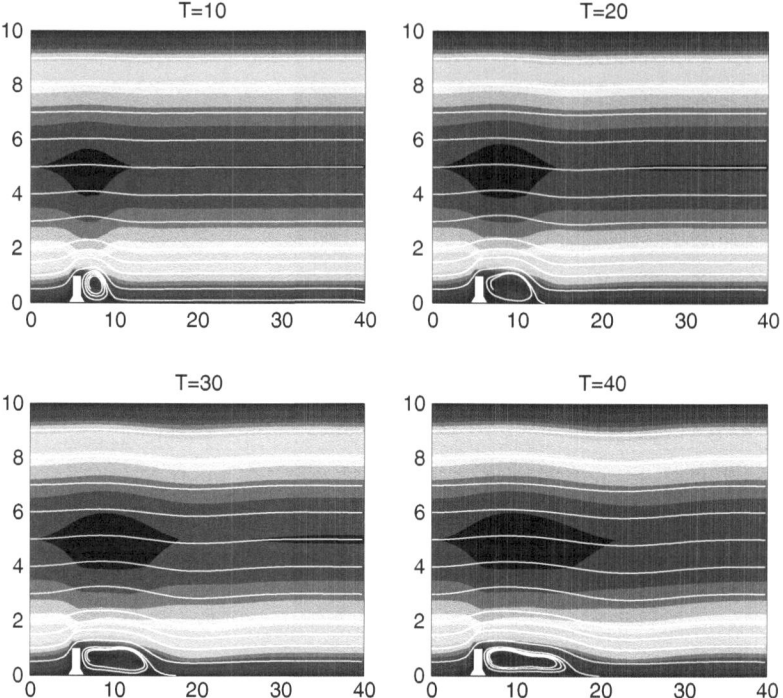

Fig. 2.6 Smagorinsky model predicts flow reaches equilibrium quickly

model predicts the flow will quickly reach a non-physical equilibrium with one long, attached eddy behind the step, Fig. 2.6.

Some modifications used to reduce the effects of eddy viscosity on the dynamics of the large structures include:

1. Taking the Smagorinsky constant small, typically $C_S \approx O(0.1)$.
2. Van Driest damping: taking $C_S = C_S(x)$, where $C_S(x) \to 0$ rapidly as $x \to \partial\Omega$.
3. The dynamic model of Germano, Piomelli, Moin and Cabot [GPMC91]: fitting $C_S(x)$ adaptively to other relations and thus even allowing for negative viscosities.

2.6 Some Smagorinsky Variants

The last experiment shows that this Smagorinsky model is clearly over diffusive. Most research on eddy viscosity models is focused on keeping just enough diffusion to represent the loss of energy due to the breakdown of eddies from resolved scales to those below the level of resolution. (There is an argument due to Lilly for picking constants in any model to fit this exact level.) Thus ideas have centered around localization: having eddy viscosity only act on places where such small eddies occur in either or both physical or wave number / scale space. We give here a few examples without much detail. Much further improvement is possible so none of these should be taken as final.

2.6.1   Using the Q-Criterion

Recall that the spin and deformation tensors, $\nabla^{ss} u$ and $\nabla^s u$, are given respectively by

$$\nabla^{ss} u := \left(\nabla u - \nabla u^{tr}\right)/2, \nabla^s u := \left(\nabla u + \nabla u^{tr}\right)/2.$$

Large, coherent, persistent vortices are marked by regions where the spin tensor dominates the deformation tensor (i.e., where rigid body rotation is stronger than deformation)

$$Q(u, u) := \nabla^{ss} u : \nabla^{ss} u - \nabla^s u : \nabla^s u > 0.$$

This condition is necessary but not sufficient condition for identifying persistent vortices in 3d and both necessary and sufficient in 2d). $Q(w, w)$ can be simplified to test:

$$Q(u, u) := \nabla u : \nabla u^{tr}.$$

Clearly, in such structures there should not be breakdown to subgrid sized eddies and accordingly no loss of energy. Thus a natural proposal (that to the author's knowledge has never been investigated) is to use $Q(u, u)$ to determine the eddy viscosity. Two examples are (where $C_S(x) \to 0$ rapidly as $x \to \partial\Omega$)

$$\nu_T := (C_S(x)\,\delta)^2 \sqrt{\max\{-Q(w, w), 0\}},$$

or, equivalently,

$$\nu_T := (C_S(x)\delta)^2 |Q(w, w)|^{1/2}, \text{ if } Q(w, w) < 0,$$
$$\nu_T := 0, \text{ if } Q(w, w) \geq 0.$$

Another option is to use it as a switch in another eddy viscosity model such as the Smagorinsky model, as in

$$\nu_T := (C_S\delta)^2 |\nabla^s w|_F, \text{ if } Q(w,w) < 0,$$

$$\nu_T := 0, \text{ if } Q(w,w) \geq 0.$$

2.6.2   A Multiscale Turbulent Diffusion Coefficient

Turbulent diffusion is caused by mixing due to the smallest resolved eddies. It thus is sensible physically to base the amount of such diffusion on the kinetic energy in them. Dimensional analysis gives the correct form (which is known as the Prandtl-Kolmogorov relation). This results in

$$\nu_T := C\,\delta\sqrt{\frac{1}{2}|w - \overline{w}|_F^2}$$

$$= C\,\delta|w - \overline{w}|_F$$

Within this theory, any similar relation that is dimensionally consistent is also acceptable. Dimensionally consistent ones are

$$\nu_T := (C_S\delta)^2 |\nabla^s(w - \overline{w})|_F$$

$$\nu_T := (C_S\delta)^3 |\triangle(w - \overline{w})|_F$$

The first of these options has been studied and tested by Hughes and his collaborators and is often called the "small-large Smagorinsky model".

2.6.3   Localization of Eddy Viscosity in Scale Space

A recent idea of Guermond (with roots in the work on spectral vanishing viscosity) and Hughes is to localize eddy viscosity in scale space so that it acts primarily on the last resolved scales. This is easiest to present variationally. The variational formulation of the induced viscous term is given by (as one option):

$$(\text{ eddy viscosity}, v) := (\nu_T\nabla^s(w'), \nabla^s(v')),$$

where $w', v' = $ fluctuations of w, v respectively.

The eddy viscosity coefficient can be chosen in many different ways resulting in various combinations and the fluctuation model can vary as well.

If fluctuations are defined by projections into finite element spaces, variational multiscale methods result. These were pioneered by Hughes and collaborators and studied in [L02] and Guermond [Guer]; see also [V03, SSK05] for its realization in non-variational numerical methods. Selecting $w' := w - \overline{w}$ the eddy viscosity operator is

$$-(I - G)\nabla \cdot [\nu_T \nabla^s((I - G)w)], G := \text{ the filter operator.}$$

This is more clear if written variationally as:

$$(\nu_T \nabla^s(w'), \nabla^s(w')) \ , \ \text{ where } w' = w - \overline{w}.$$

As a concrete example, Hughes calls the following the "small-small Smagorinsky" model

$$\cdots - (I - G)\nabla \cdot [(C_S\delta)^2 |\nabla^s((I - G)w)|_F \nabla^s((I - G)w)]$$

or, variationally,

$$\cdots + ((C_S\delta)^2 |\nabla^s w'|_F \nabla^s w', \nabla^s w')$$

and has shown that it has accuracy for statistics of turbulent flows comparable to much more complex and expensive models. This idea has great promise because it results in eddy viscosity operators that can be localized in both physical space and scale space, exactly as intended by Boussinesq!

2.6.4 Vreman's Eddy Viscosity

Perhaps the most advanced eddy viscosity coefficient has recently been proposed by Bert Vreman in [V04]. In a deep construction, using only the gradient tensor he constructs an eddy viscosity coefficient formula that vanishes identically for 320 types of flow structures that are known to be coherent (non turbulent). The full justification for the formula is given in his paper. Briefly, it is given as follows

$$\nu_T = C\delta^2 \sqrt{\frac{B}{|\nabla w|_F^4}}, \ \text{ and if the denominator vanishes } \nu_T = 0,$$

where

$$\beta_{ij} := \sum_{m=1,2,3} \frac{\partial w_i}{\partial x_m} \frac{\partial w_j}{\partial x_m}$$

$$B(\beta) := \beta_{11}\beta_{22} - \beta_{12}^2 + \beta_{11}\beta_{33} - \beta_{13}^2 + \beta_{22}\beta_{33} - \beta_{23}^2$$

2.7   A Glimpse into Near Wall Models

> "It is disappointing to find that the boundary regions in large-eddy simulations contain serious errors. This cannot, however, be considered too surprising, as close to the surface the potential rationality of the large-eddy simulation vanishes as the dominant eddy-scales become comparable with, and smaller than, the filter-scale."
> P. J. Mason, in: *Large eddy simulation: a critical review of the technique*, Q.J.R. Meteorol. Soc. 120 (1994), 1–26.

> "Always present is the temptation to look for the key under the streetlight and not in the dark corner where it was lost. ... Much of the extensive turbulence research today follows this line and fails to address the real problems."
> H. W. Liepmann, 1997.

Typical flow geometries in turbulent flow have inflow boundaries, obstacles, walls and outflow boundaries and correct boundary conditions must be imposed for all of them. Even the simplest case (which we are considering) of an internal flow problem is already not so simple, as we shall see! Let $\bar{u} = g_\delta * u$, $\bar{p} = g_\delta * p$ and let (w, q) approximate $(\bar{u}, \bar{p})$ and satisfy the initial value problem:

$$w_t + \nabla \cdot (w\, w) + \nabla q - Re^{-1}\triangle w - \nabla \cdot (\nu_T \boldsymbol{\nabla}^s w) = \bar{f}, \text{ in } \Omega \times (0, T],$$
$$\nabla \cdot w = 0, \ in \ \Omega \times (0, T], \text{ and } w(x, 0) = \bar{u}_0(x), \text{ in } \Omega.$$

In practical calculations, the most commonly used boundary condition for w is simply

$$w = 0, \text{ on } \partial\Omega. \tag{2.23}$$

Recall that $w \cong \bar{u}$. Consider Fig. 2.7; it is clear that imposing the no slip condition (2.23) on the large eddies introduces a consistency problem since $\bar{u} = g_\delta * u \neq 0$ on $\partial\Omega$: $u = 0$ on $\partial\Omega$ but $\bar{u}$ being an $O(\delta)$ average of u, does *not* vanish on $\partial\Omega$.

It is also clear that large eddies do not stick to boundaries if you consider common examples like tornadoes and hurricanes. When contacting the earth, these do move along the earth and lose energy as they move. The picture painted by boundary layer theory and these last examples give important clues to finding the correct boundary conditions to be *no-penetration* and *slip with friction*. Mathematically, these are written as:

$$\begin{cases} w \cdot \hat{n} = 0, \ on \ \partial\Omega, \ and \\ w \cdot \hat{\tau}_j + \beta(\delta, Re)\hat{n} \cdot \boldsymbol{\Pi}(w) \cdot \hat{\tau}_j = 0 \ on \ \partial\Omega, \end{cases} \tag{2.24}$$

where $(\hat{n}, \hat{\tau}_j)$ are the unit tangent and normal vectors and $\boldsymbol{\Pi}(w)$ is the stress tensor associated with the viscous and turbulent stresses:

$$\hat{n} \cdot \boldsymbol{\Pi} \cdot \hat{\tau}_j = \hat{n} \cdot (2Re^{-1}\nabla^s w + \nu_T \nabla^s w) \cdot \hat{\tau}_j.$$

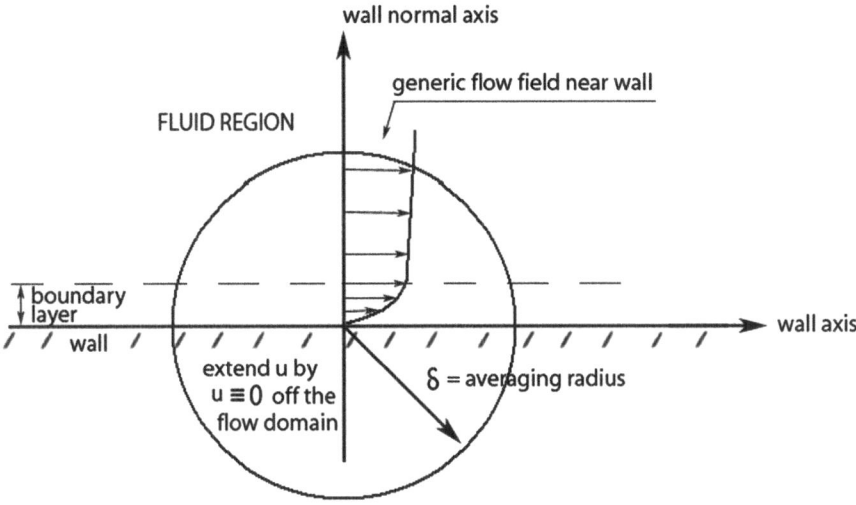

Fig. 2.7 $\bar{u}$ does *not* vanish on $\partial\Omega$

The modeling problem now reduces to identification of the (linear or even nonlinear) friction coefficient β. The early analysis of Maxwell gives some insight into the friction coefficient. If we identify the micro-length scale with the filter length scale δ, then his derivation of the no slip condition (reviewed in [L08]) suggests

$$\beta \sim Re\frac{\delta}{L}.$$

With this scaling, as $\delta \to 0$ we recover the no-slip condition and as $Re \to \infty$ we transition to free-slip and the Euler equations. Thus, this passes the test of asymptotic reasonableness. Of course, we do not compute in either asymptotic regime so a good formula must be found for the friction coefficient in the intermediate region. For initial attempts, see [GL00] and [JLS04].

2.8 Remarks

There are several excellent books connecting mathematics and basic physical understanding of turbulence, including the books of Frisch [F95] and Pope [Po00]. There are many experimental studies of turbulence pointing to the (still unproven, although progress continues to be made [F97], but widely believed to be true) idea that the large scales evolve deterministically while sensitive dependence is restricted to the small turbulent fluctuations.

The overview of LES given in this chapter is fairly standard. For good books on large eddy simulation, see the books of Sagaut [S01], Geurts [G03], John [J04] and Berselli, Iliescu and WL [BIL06]. Numerical analysis of the Smagorinsky model was performed in [JL00], by Pares in [P92, P94] and Coletti [C98]. It is related to work on the Ladyzhenskaya model of Du and Gunzburger [DG91]. There are many eddy viscosity models, e.g., [IL98, LL02, D03, GP05]; most are derived to improve various features of the Smagorinsky model and are compared with it in computational tests. For example, the work on (the very promising approach of) variational multiscale methods is large and growing fast, see, for example, [HKJ00, HOM01, LL02, LLe06, Guer, BBJL07]; its realization in non-variational numerical methods also works very well, e.g., [V03, SSK05].

The connection of filtering to scale space analysis is from [LMNR09]. Applying scale space analysis systematically to LES filtering is a promising and yet open area. The outline to one approach to wall laws is developed in [DJL04], see also [JL06, L99, Li01, BFG02] and [Day90, Max79] for some history. There are many other approaches to generating effective boundary conditions for the large eddies. Finding good near wall models, including addressing the commutator error problem [DJL04, BJG07], that provide accuracy in the mean flow without resolving the boundary layers is a very important and difficult problem in CFD.

Chapter 3
Approximate Deconvolution Operators and Models

3.1 Useful Deconvolution Operators

The great challenge in simulation of turbulent flows from applications ranging from geophysics to biomedical device design is that equations for the pointwise flow quantities are well-known but intractable to computational solution and sensitive to uncertainties and perturbation in problem data. On the other hand, closed equations for the averages of flow quantities cannot be obtained directly from the physics of fluid motion. Thus, modeling in large eddy simulation (meaning the approximation of local, spacial averages in a turbulent flow) is typically based on guesswork (phenomenology), calibration (data fitting model parameters) and (at best) approximation.

The goal of large eddy simulation is thus to predict reliably the evolution of large scales of turbulent flows. Thus, LES is inherently concerned with dynamics of turbulence. The definition of *large scales* is done through a local, spacial averaging process associated with the chosen averaging radius δ. Three types of averaging processes are common in large eddy simulations: filtering by convolution, differential filters and filtering by projection.

Once an averaging radius and a filtering process is selected, an LES model must be developed. Broadly, this chapter is concerned with approximate deconvolution models. Averaging the NSE leads to the (non-closed) Space Filtered Navier–Stokes Equations (SFNSE) for $\overline{u}$, $\overline{p}$:

$$\overline{u}_t + \nabla \cdot \overline{uu} - \nu \Delta \overline{u} + \nabla \overline{p} = \overline{f}, \text{ and } \nabla \cdot \overline{u} = 0. \tag{3.1}$$

In approximate deconvolution models one constructs an approximate filter inverse

$$D := \text{approximate filter inverse}$$

W.J. Layton and L.G. Rebholz, *Approximate Deconvolution Models of Turbulence*, Lecture Notes in Mathematics 2042, DOI 10.1007/978-3-642-24409-4_3, © Springer-Verlag Berlin Heidelberg 2012

then the model for the non closed terms is

$$\overline{uu} \cong \overline{D(\overline{u})D(\overline{u})}.$$

Calling w the resulting approximation to $\overline{u}$, the basic ADM is then

$$w_t + \nabla \cdot \overline{D(w)D(w)} - \nu \Delta w + \nabla q = \overline{f}, \ and \ \nabla \cdot w = 0. \qquad (3.2)$$

In the simplest (and least accurate) $N = 0$ case, the operator $D = I$ so $D_0\overline{u} = \overline{u}$ and thus the Zeroth Order ADM is

$$w_t + \nabla \cdot (\overline{w}\overline{w}) - \nu \Delta w + \nabla q = \overline{f}, \ \nabla \cdot w = 0. \qquad (3.3)$$

This model has a unique, strong solution and the smoothness of this solution is limited only by the smoothness of the problem data u_0 and f. The analysis is quite clear for the zeroth order ADM and the *ideas* in the general case are the same as for zeroth order ADM.

Once a model is selected, important and fundamental questions arise concerning whether the model correctly captures the global energy balance of the large scales. Mathematically, this is expressed as the question.

- *Does the LES model have unique, strong solutions which depend stably on the problem data?*

Broadly, there are two types of LES models of turbulence: *descriptive* or phenomenological models (e.g., eddy viscosity models) and *predictive* models (considered herein). The accuracy of a model measured in a chosen norm, $||\cdot||$, means

$$||average(NSE \ solution) - LES \ solution||.$$

A model's accuracy can be assessed in several experimental and analytical ways. The strongest test is to obtain a Navier–Stokes equation solution from a data base, a DNS or experimental data from a real flow, explicitly average it and compare it to the velocity predicted by a large eddy simulation.[1] Another important experimental approach is to use a velocity field from a direct numerical simulation (a DNS) to compute some norm of the model's *consistency error* (defined precisely below) evaluated at the DNS solution (which is regarded as a "truth solution").[2] Either experimental

---

[1] Because of the uncertainties and fluctuations of turbulent flow, often (or usually) "compare" means to compare time-averaged flow statistics rather than compute the time evolution of norms of differences.

[2] These two approaches are known in LES as á priori testing and a posteriori testing of the model under consideration.

approach of course limits assessments to Reynolds numbers and geometries for which reliable DNS data is available. A third complementary approach (for which there are currently few results) is to study analytically the model's *consistency error* as a function of the averaging radius δ and the Reynolds number Re. The inherent difficulties in analytical studies are that:

- *Consistency error bounds for infinitely smooth functions hardly address essential features of turbulent flows such as irregularity and richness of scales.*
- *Worst case bounds for general weak solutions of the Navier Stokes equations are so pessimistic that they yield little insight.*

However, it is known that after time or ensemble averaging, turbulent velocity fields are often observed to have an intermediate smoothness as predicted by the Kolmogorov theory (often called the K41 theory). This case is often referred to as homogeneous isotropic turbulence and various norms of flow quantities can be estimated by using it, Parseval's equality and spectral integration. We mentioned Lilly's famous paper as an early and important example.

3.1.1 Approximate Deconvolution

First, we note that with a differential filter, exact deconvolution is possible but not useful (an observation of M. Germano). Indeed, since $\overline{\phi} = (-\delta^2\triangle + 1)^{-1}\phi$, the solution of the deconvolution problem is $\phi = (-\delta^2\triangle + 1)\overline{\phi}$. Let the operator A denote this exact deconvolution operator $A := (-\delta^2\triangle + 1)$. Clearly A is not a bounded operator on $L^2(\Omega)$ so exact deconvolution is unstable to high frequency perturbations (this is known as the small divisor problem). Further, if exact deconvolution is used to obtain the (exact) equation for the flow averages, we obtain

$$\overline{u}_t + \nabla \cdot (\overline{A\overline{u}\ A\overline{u}}) - \nu\triangle\overline{u} + \nabla\overline{p} = \overline{f}, \text{ and } \nabla \cdot \overline{u} = 0.$$

The important observation is that this is just a change of variables in the Navier–Stokes equations. Thus, it does not reduce the amount of information contained in the model's solution and any theory of the model will share all the gaps in its mathematical theory with that of the Navier–Stokes equations. Thus, only approximate deconvolution can produce a useful model.

The basic problem in approximate deconvolution is thus: given $\overline{u}$ find *useful* approximations of u. In other words, solve the equation

$$Gu = \overline{u}, \text{ solve for } u.$$

For most averaging operators, G is symmetric and positive semi-definite. Typically, G is not invertible or at least not stably invertible due to small divisor problems. Thus, this deconvolution problem is ill-posed. Also typically, G is invertible on a subspace which becomes dense in $L^2(\mathbb{R}^3)$ as $\delta \to 0$.

The filtering or convolution operator $u \to \overline{u}$ is a bounded map: $L^2(\Omega) \to L^2(\Omega)$. If (as in the case we study) it is smoothing, its inverse cannot be bounded due to small divisor problems. An approximate deconvolution operator D_N is an approximate inverse $\overline{u} \to D_N(\overline{u}) \approx u$ which:

- Is a bounded operator on $L^2(\Omega)$.
- Approximates u in some useful (typically asymptotic) sense.
- Satisfies other conditions necessary for the application at hand.

The deconvolution operator we develop herein was studied by van Cittert in 1931, e.g., Bertero and Boccacci [BB98], and we note that inverse modeling in LES was studied previously in [Geu97]. The use of deconvolution in LES was pioneered by Stolz, Adams and Kleiser [AS01, SA99, AS02, SAK01a, SAK01b, SAK02]. The Nth van Cittert approximate deconvolution operator D_N is defined by N steps of Picard iteration, [BB98]:

given $\overline{u}$ solve $Gu = \overline{u}$ for u.

by N steps of: $u_0 = \overline{u}$ and $u_{new} = u_{old} + \{\overline{u} - Gu_{old}\}$

The Nth van Cittert deconvolution operator D_N is given explicitly by

$$D_N\phi := \sum_{n=0}^{N}(I - G)^n\phi. \tag{3.4}$$

Definition 12. Given a velocity field $u(x,t)$, the deconvolution error at time t and time averaged deconvolution errors are, respectively,

$$||u(\cdot,t) - D\overline{u}(\cdot,t)||, \quad \text{and} \quad \left\langle ||u - D\overline{u}||^2 \right\rangle^{1/2}.$$

The deconvolution error $||u - D_N\overline{u}||$ in van Cittert deconvolution for a C^∞ velocity is $O(\delta^{2N+2})$. However, turbulent velocities are not infinitely smooth (uniformly in the Reynolds number). In [LL05] the following was proven about the deconvolution error for turbulent velocities with a typical $k^{-5/3}$ energy spectrum predicting slow but uniform in Re convergence.

Theorem 13. *Let u be a turbulent velocity with time averaged energy spectrum $E(k)$ satisfying $E(k) \leq \alpha\varepsilon^{2/3}k^{-5/3}$. The time averaged deconvolution error of van Cittert deconvolution satisfies*

$$\left\langle ||u - D_N\overline{u}||^2 \right\rangle \leq \left(3 + \frac{2}{4N + 10/3}\right)\alpha\varepsilon^{2/3}\left(\frac{\delta}{L}\right)^{2/3}.$$

3.2 LES Approximate Deconvolution Models

Averaging/filtering the NSE leads to the (non-closed) Space Filtered Navier–Stokes Equations (SFNSE) for $\overline{u}$, $\overline{p}$:

$$\overline{u}_t + \nabla \cdot \overline{uu} - \nu \Delta \overline{u} + \nabla \overline{p} = \overline{f}, \text{ and } \nabla \cdot \overline{u} = 0. \tag{3.5}$$

An LES model arises by replacing the (non-closed) tensor $\overline{uu}$ by one depending only on $\overline{u}$, and not u. The action of the resulting LES model should:

A1. Be very close to the SFNSE on the large scales (or smooth flow components).
A2. Truncate scales: the solution of the approximate model for $\overline{u}$ should have fewer persistent scales than u.
A3. Give an accurate representation of the mean effects of the unresolved scales on the resolved scales.

These three conditions are common in the LES literature and an interpretation of them plays a key role in the models studied herein. The (ill-posed) deconvolution problem is:

$$\text{given } \overline{u}, \text{ find } u \text{ approximately}$$

Let a selected approximate deconvolution of $\overline{u}$ be denoted

$$\overline{u} \rightarrow D(\overline{u}) := \text{approximation to } u.$$

Then the closure problem can be solved (approximately) by

$$\overline{uu} \approx \overline{D(\overline{u})D(\overline{u})}$$

and the induced LES model is $\nabla \cdot w = 0$ and

$$w_t + \nabla \cdot (\overline{D(w)D(w)}) - \nu \Delta w + \nabla q = \overline{f}. \tag{3.6}$$

For comparison, the *exact* SFNSE can be rewritten as $\nabla \cdot \overline{u} = 0$ and

$$\overline{u}_t + \nabla \cdot (\overline{D(\overline{u})D(\overline{u})}) - \nu \Delta \overline{u} + \nabla \overline{p} = \overline{f} + \nabla \cdot (\overline{D(\overline{u})D(\overline{u}) - uu}). \tag{3.7}$$

Comparing the exact SFNSE with the LES model, the difference is the tensor $D(\overline{u})D(\overline{u}) - uu$ on the right hand side. Comparing, the condition (A1) that the model is very close to the exact SFNSE for the largest scales can be interpreted that

$$\phi \rightarrow \overline{D(\overline{\phi})D(\overline{\phi})} - \overline{\phi\phi}$$

is very small for ϕ the smooth components (or largest scales) of u. Since

$$D(\overline{u})D(\overline{u}) - uu = D(\overline{u})(D(\overline{u}) - u) + u(D(\overline{u}) - u),$$

it clearly suffices for (A1) that the deconvolution error $\phi - D(\overline{\phi})$ be small for smooth ϕ.

The scale truncation condition (A2) can be traced back to a condition on the deconvolution operator D as well. If the averaging operator is smoothing, then *exact* deconvolution must be an unbounded operator because it must increase the energy in the small scales / high frequencies (exactly as much as the averaging operator decreases it). The intuition is thus that *truncation* in the LES model is derived by a *bounded* approximate deconvolution operator D (approximating exact deconvolution well on the low frequencies).

The third condition (A3), that the model is a faithful representation of the mean effects of the unresolved scales on the resolved scales, is a deeper condition related to the physical fidelity of the flow statistics predicted by the model. One way to obtain useful insight into (A3) was made explicit by Muschinsky [Mus96] for the Smagorinsky model. In this approach, the conditions necessary for solutions of an LES model to exhibit an energy cascade are checked. Next, a similarity theory of the model is developed and the turbulent statistics predicted by the model are compared to those of the Navier–Stokes equations through the filter length scale / cutoff frequency. Each such test of a model gives evidence that, in the mean, the representation of the mean effects of the unresolved scales is accurate and that important flow statistics are correctly predicted.

3.3 Examples of Approximate Deconvolution Operators

The filtering or convolution operator $G : u \to \overline{u}$ is a bounded map: $L^2(\Omega) \to L^2(\Omega)$. If (as in the case we study) it is smoothing, its inverse cannot be bounded due to small divisor problems. Indeed, it is known quite generally that inversion is not well posed. Consider the problem: given $\overline{u} + noise$

$$Gu = \overline{u} \ (+noise), \text{ solve for } u.$$

Theorem 14. *Let H be a Hilbert space and $G : H \to H$ a compact map. Then, if H is infinite dimensional*

$$Range(G) \neq H.$$

In other words, G is not invertible as a bounded linear operator.

Thus stable exact deconvolution is not possible and approximate deconvolution must be used instead. An approximate deconvolution operator D_N is an approximate inverse $\overline{u} \rightarrow D_N(\overline{u}) \approx u$ which

- Is a bounded operator on $L^2(\Omega)$
- Approximates u in some useful (typically asymptotic) sense
- Satisfies other conditions necessary for the application at hand

The most common and universal solution to the deconvolution problem is Tikhonov regularization.

3.3.1 Tikhonov Regularization

A small parameter μ is selected. Given $\overline{u}$ an approximation to u is determined as the solution of the minimization problem:

$$u_\mu := \text{minimizer over } L^2(\Omega) \text{ of } J_\mu(v)$$

$$J_\mu(v) := \frac{1}{2}||G^*Gv - G^*\overline{u}||^2 + \frac{\mu}{2}||v||^2.$$

The Euler-Lagrange equations of this minimization problem yield

$$u_\mu = (G^*G + \mu I)^{-1}G^*\overline{u}.$$

This gives the approximate deconvolution operator

$$D_\mu = (G^*G + \mu I)^{-1}G^*.$$

3.3.2 Tikhonov-Lavrentiev Regularization

If the filtering operator is self-adjoint and positive definite (the typical case) then the minimization problem can be modified as follows. Given $\overline{u}$ an approximation to u is determined as the minimizer of:

$$J_\mu(v) := \frac{1}{2}(Gv, v) - (\overline{u}, v) + \frac{\mu}{2}||v||^2.$$

The Euler-Lagrange equations of this minimization problem yields

$$u_\mu = (G + \mu I)^{-1}\overline{u},$$

and the approximate deconvolution operator

$$D_\mu = (G + \mu I)^{-1}.$$

3.3.3   A Variant on Tikhonov-Lavrentiev Regularization

Manica and Stanculescu have shown in [MS09] that the accuracy of Tikhonov-Lavrentiev regularization for deconvolution problems can be increased from $O(\mu)$ to $O(\mu\delta^2)$ by a very simple modification. Given $\overline{u}$ an approximation to u is determined as the minimizer of:

$$J_\mu(v) := (1 - \mu) \left[\frac{1}{2}(Gv, v) - (\overline{u}, v) \right] + \frac{\mu}{2}||v||^2.$$

The Euler-Lagrange equations of this minimization problem yields

$$u_\mu = ((1 - \mu)G + \mu I)^{-1}\overline{u},$$

and the approximate deconvolution operator

$$D_\mu = ((1 - \mu)G + \mu I)^{-1}.$$

3.3.4   The van Cittert Regularization

The Nth van Cittert approximate deconvolution operator $D_N\overline{u} := \sum_{n=0}^{N}(I - G)^n\overline{u}$, introduced earlier, is defined by N steps of repeated filtering, [BB98], given $u^{old} = \overline{u}$:

$$N - steps : u^{new} = u^{old} + \{\overline{u} - Gu^{old}\}$$

Clearly, this is nothing but the first order Richardson iteration for solving the operator equation $Gu = \overline{u}$ involving a possibly non-invertible operator G. Since the deconvolution problem is ill-posed, convergence as $N \to \infty$ cannot expected. The relevant question is convergence for fixed N as $\delta \to 0$.

3.3.5   van Cittert with Relaxation Parameters

Since the van Cittert deconvolution method is nothing but first order Richardson, relaxation parameters can be introduced into the deconvolution procedure without any increase in computational effort.

Algorithm 15 (van Cittert with relaxation). *Given relaxation parameters $\tau_n, n = 0, \cdots, N$, set $u_0 = \overline{u}$,*
   For $n = 1, 2, \ldots, N - 1$, perform: $u_{n+1} = u_n + \tau_n\{\overline{u} - Gu_n\}$

This leads to the optimization problem:

Pick the relaxation parameters τ_n to minimize the time averaged deconvolution error $\langle ||u - D\overline{u}||^2 \rangle^{1/2}$ for all velocity fields satisfying a given structural condition.

Two parameter selection choices have been considered so far[3]: picking the method parameters to optimize over (1) all velocities possessing a K41 type energy spectrum and (2) all velocities having finite kinetic energy. For example, for ones with finite kinetic energy the parameters are given (through a change of variable related to the Chebychev points) by

$$\tau_n = \left(\frac{\pi}{2}\right)^2 \left(1 + \cos\left[\frac{2n+1}{4N}\pi\right]\right)^2, n = 1, 2, \cdots, N.$$

In [LS09], the following was proven about the deconvolution error for turbulent velocities with a typical $k^{-5/3}$ energy spectrum predicting slow but uniform in Re convergence.

Theorem 16. *Let u be a turbulent velocity with time averaged energy spectrum $E(k)$ satisfying $E(k) \leq \alpha\varepsilon^{2/3}k^{-5/3}$. Let the relaxation parameters be given by*

$$\tau_n = \left(\frac{\pi}{2}\right)^2 \left(1 + \cos\left[\frac{2n+1}{4N}\pi\right]\right)^2, n = 1, 2, \cdots, N.$$

The time averaged deconvolution error of van Cittert deconvolution with relaxation satisfies

$$\langle ||u - D_N\overline{u}||^2 \rangle \leq 2 \left(\frac{\pi}{4}\right)^N \alpha\varepsilon^{2/3} \left(\frac{\delta}{L}\right)^{2/3}.$$

3.3.6 Other Approximate Deconvolution Methods

The development of approximate deconvolution models in LES does not seem to have progressed much further beyond the van Cittert ADM without parameters. Thus it is natural to consider more sophisticated algorithms (which are still computationally simple) such as second order Richardson.

Algorithm 17 (Second order Richardson deconvolution). *Given relaxation parameters τ_n, set $u_0 = \overline{u}$, and $u_1 = u_0 + \tau_0\{\overline{u} - Gu_0\}$*

---

[3]The case of picking the relaxation parameters to optimize deconvolution errors inside a turbulent boundary layer is an important open problem.

For $n = 1, 2, \ldots, N$, perform
$$\tilde{u}_{n+1} = u_n + \tau_n\{\overline{u} - Gu_n\} \quad and \quad u_{n+1} = \theta_n \tilde{u}_{n+1} + (1 - \theta_n)u_{n-1}$$

Other algorithms are open for analysis and testing as well, such as steepest descent and the conjugate gradient method. It is important to note though that these produce approximations to u which depend nonlinearly on $\overline{u}$. Thus both result in an LES model with more complex nonlinearities than the SFNSE. Basic questions like existence and uniqueness of the resulting model's solutions become much more complex and difficult.

3.4   Analysis of van Cittert Deconvolution

We study the properties of the N*th* van Cittert approximate deconvolution operator D_N, given by

$$D_N\phi := \sum_{n=0}^{N}(I - G)^n\phi. \tag{3.8}$$

Lemma 18 (Stability of approximate deconvolution). *Let G be a bounded, self-adjoint, positive operator with $0 \leq \lambda(G) \leq 1$. Let $||\cdot||$ denote the $L^2(\Omega)$ operator norm. Then D_N is a self-adjoint, positive semi-definite operator on $L^2(\Omega)$ and*

$$||D_N|| \leq N + 1.$$

Proof. First, note that since G is a self-adjoint positive definite operator with eigenvalues between zero and one and D_N is a function of G, D_N is also self-adjoint. By the spectral mapping theorem

$$\lambda(D_N) = \sum_{n=0}^{N}\lambda(I - G)^n = \sum_{n=0}^{N}(I - \lambda(G))^n.$$

Thus, the eigenvalues of D_N are non-negative and D_N is also positive semi-definite. Since D_N is self-adjoint, the operator norm $||D_N||$ is also easily bounded by the spectral mapping theorem by

$$||D_N|| = \sum_{n=0}^{N}\lambda_{\max}(I - G)^n = \sum_{n=0}^{N}(1 - \lambda_{\min}(G))^n = N + 1. \tag{3.9}$$

$\square$

Definition 19. The deconvolution weighted inner product and norm, $(\cdot, \cdot)_N$ and $||\cdot||_N$ are

$$(u, v)_N := (u, D_N v), \quad ||u||_N := (u, u)_N^{\frac{1}{2}}$$

Lemma 20. *Let G be a bounded, self-adjoint, positive operator. Consider the van Cittert approximate deconvolution operator*

$$D_N : L^2(\Omega) \to L^2(\Omega).$$

D_N is a bounded, self-adjoint, positive-definite operator. If $0 \leq \lambda(G) \leq 1$, the norm $|| \cdot ||_N$ is uniformly in δ equivalent to the usual L^2 norm. It satisfies

$$||\phi||^2 \leq ||\phi||_N^2 \leq (N+1)||\phi||^2, \ \forall \phi \in L^2(\Omega).$$

Proof. D_N is a function of the bounded, self-adjoint operator G and is thus bounded and self-adjoint. By the spectral mapping theorem we have

$$\lambda(D_N) = \sum_{n=0}^{N} \lambda(I - G)^n = \sum_{n=0}^{N}(1 - \lambda(G))^n, \text{ and}$$

$$0 < \lambda(G) \leq 1.$$

Thus, $1 \leq \lambda(D_N) \leq N+1$. Since D_N is a self-adjoint operator, this proves positive definiteness and the above equivalence of norms.   □

In the periodic case the condition that $0 \leq \lambda(G) \leq 1$ is equivalent to the condition that $0 \leq \widehat{G}(k) \leq 1$. This can be verified easily by a Fourier series calculation. For example, for *differential filters* we have

$$\widehat{G}(k) = \frac{1}{\delta^2|k|^2 + 1},$$

while for the *Gaussian filter*

$$\widehat{G}(k) = e^{-\delta^2|k|^2}.$$

Both are bounded between zero and one, as assumed. Since van Cittert gives deconvolution operators as a truncated geometric series, its error (in that truncation) is easy to estimate.

Lemma 21 (Error in approximate deconvolution). *Let the averaging be defined by the differential filter $G\phi = (-\delta^2 \triangle + 1)^{-1}\phi$ subject to periodic boundary conditions. Let $A = -\delta^2 \triangle + 1$. For any $\phi \in L^2(\Omega)$,*

$$\phi - D_N \overline{\phi} = (I - G)^{N+1}\phi$$

$$= (-1)^{N+1}\delta^{2N+2} \triangle^{N+1} A^{-(N+1)}\phi.$$

Proof. Let $B = I - G$. Since $\overline{\phi} = G\phi, \overline{\phi} = (I - B)\phi.$

Since $D_N := \sum_{n=0}^{N} B^n$, a geometric series calculation gives

$$(I - B)D_N\overline{\phi} = (I - B^{N+1})\overline{\phi}.$$

Subtraction gives

$$\phi - D_N\phi = AB^{N+1}\overline{\phi} = B^{N+1}A\overline{\phi} = B^{N+1}\phi.$$

Finally, $B = I - G$, so rearranging terms gives

$$\phi - D_N\phi = (A - I)^{N+1}A^{-(N+1)}\phi$$
$$= A^{-(N+1)}((-1)^{N+1}\delta^{2N+2}\triangle^{N+1})\phi,$$

which are the claimed results.   □

It is insightful to consider the Cauchy problem or the periodic problem and visualize the approximate deconvolution operators D_N in wave number space (re-scaled by $k \leftarrow \delta k$). For G a differential filter, the transfer function or symbol of the first three are

$$\widehat{G_0} = 1,$$
$$\widehat{G_1} = 2 - \frac{1}{k^2 + 1} = \frac{2k^2 + 1}{k^2 + 1}, \text{ and}$$
$$\widehat{G_2} = 1 + \frac{k^2}{k^2 + 1} + \left(\frac{k^2}{k^2 + 1}\right)^2.$$

These three are plotted in Fig. 3.1 together with the transfer function of exact deconvolution. The transfer function of exact deconvolution of a differential filter is $k^2 + 1$. The next figure plots this in bold against the above approximate deconvolution operators.

The large scales are associated with the smooth components and with the wave numbers near zero (i.e., $|k|$ small). Thus, the fact that D_N is a very accurate solution of the deconvolution problem for the large scales is reflected in the above graph in that the transfer functions have high order contact near $k = 0$.

The approximate deconvolution operator D_N is a *bounded* operator which approximates the *unbounded* exact deconvolution operator to high asymptotic accuracy $O(\delta^{2N+2})$ on subspaces of smooth functions. In this section we restrict our analysis to the periodic problem and averaging by differential filters. It is insightful visualize the approximate deconvolution operators D_N in terms of the transfers function of the operator $\widehat{D_N}$. Since these are functions of δk, where $k = |\mathbf{k}|$, and not δ or $\mathbf{k}$, it is appropriate to record them *re-scaled by*

$$k \leftarrow \delta k.$$

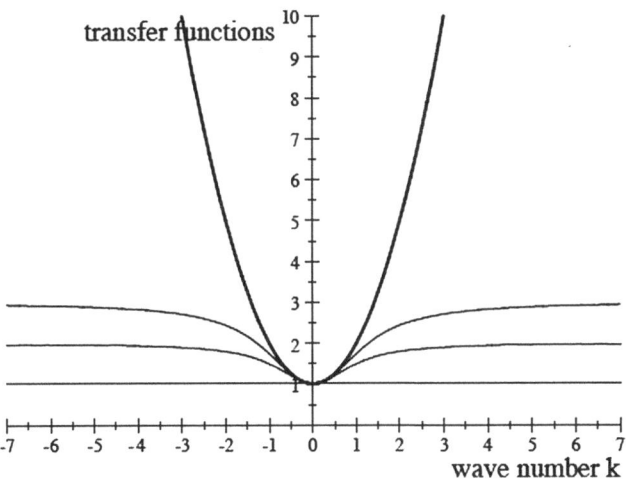

Fig. 3.1 Shown above, from top to bottom, are the exact and $N = 0, 1, 2$ approximate deconvolution operators

Since $G = (-\delta^2 \triangle + 1)^{-1}$, we find, after rescaling, $\widehat{G}(k) = \frac{1}{1+k^2}$. With that, the transfer function of the first three deconvolution operators are

$$\widehat{D_0} = 1,$$

$$\widehat{D_1} = 2 - \frac{1}{k^2 + 1} = \frac{2k^2 + 1}{k^2 + 1}, \text{ and}$$

$$\widehat{D_2} = 1 + \frac{k^2}{k^2 + 1} + \left(\frac{k^2}{k^2 + 1}\right)^2.$$

More generally, we find

$$\widehat{D_N}(k) := \sum_{n=0}^{N} \left(\frac{1}{1+k^2}\right)^n = (1+k^2)\left[1 - \left(\frac{k^2}{1+k^2}\right)^{N+1}\right].$$

The large scales are associated with the wave numbers near zero (i.e., $|k|$ small). Thus, the fact that D_N is a very accurate solution of the deconvolution problem for the large scales is reflected in Fig. 3.1 in that the transfer functions have high order contact near $k = 0$. The key observation in studying asymptotics of the model as $N \to \infty$ is that (loosely speaking) the region of high accuracy grows (slowly) as N increases.

The regularization in the nonlinearity involves a special combination of averaging and deconvolution $D_N G : L^2(\Omega) \to L^2(\Omega)$ that acts like a spectral truncation operator.

Proposition 22. *For each* $N = 0, 1, \cdots$, *the operator* $D_N G$ *is positive semi-definite on* $L^2(\Omega)$. *The operator* $D_N G : L^2(\Omega) \to L^2(\Omega)$ *is a compact operator. Moreover,* $D_N G$ *maps continuously* $L_0^2(\Omega)$ *into* H^2. *Further,*

$$||D_N G||_{\mathcal{L}(L^2(\Omega) \to L^2(\Omega))} = 1, \text{ for each } N \geq 0,$$

$$||D_N G||_{\mathcal{L}(H^1 \to H^1)} = 1, \text{ for each } N \geq 0.$$

Proof. These properties are easily read off from the transfer function $\widehat{D_N G}(k)$ of $D_N G$ which we give next. For example, compactness follows since $|\widehat{D_N G}(k)| \to 0$ as $|k| \to \infty$. $\square$

The Fourier coefficients/transfer function of the operator $D_N G$ are similarly easily calculated to be (after rescaling by $k \leftarrow \delta k$)

$$\widehat{D_N G}(k) = 1 - \left(\frac{k^2}{1 + k^2} \right)^{N+1}. \tag{3.10}$$

The plots, shown in Fig. 3.2, are representative of the behavior of the whole family. Examining the above graphs, we observe that $D_N G(u)$ is very close to u for the low frequencies/largest solution scales and that $D_N G(u)$ attenuates small scales/high frequencies. The breakpoint between the low frequencies and high frequencies is somewhat arbitrary. The following is convenient for our purposes and fits our intuition of an approximate spectral cutoff frequency. We take for k_c the frequency for which $\widehat{D_N G}$ most closely attains the value $\frac{1}{2}$.

Definition 23 (Cutoff-Frequency). The cutoff frequency of $D_N G$ is

$$k_c := \text{greatest integer} \left(\widehat{D_N G}^{-1} \left(\frac{1}{2} \right) \right).$$

That is, the frequency for which $\widehat{D_N G}$ most closely attains the value $\frac{1}{2}$.

From the above explicit formulas, it is easy to verify that the cutoff frequency grows to infinity slowly as $N \to \infty$ for fixed δ and as $\delta \to 0$ for fixed N. Other properties (whose proofs are simple calculations) of the operator $D_N G(\cdot)$ follow easily from $\widehat{D_N G}(k) = 1 - \left(\frac{k^2}{1+k^2} \right)^{N+1}$.

Lemma 24. *For all* $N \geq 0$, $\widehat{D_N G}(k)$ *has the following properties:*

$$0 < \widehat{D_N G}(k) \leq 1,$$

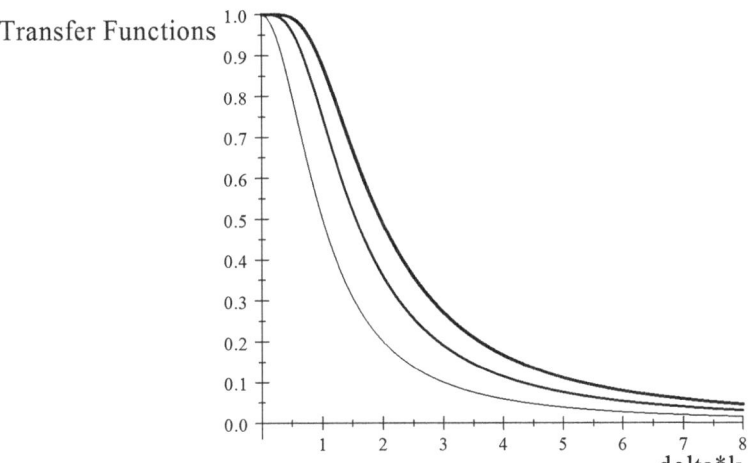

Fig. 3.2 Transfer functions DG, N=0,1,2

$$|\widehat{D_N G}(k)| \to 0 \ as \ |k| \to \infty,$$

$$\widehat{D_N G}(0) = 1.$$

Let k_c be cutoff frequency, then

$$|k_c| \to \infty \ as \ N \to \infty \ for \ fixed \ \delta \ and \ as \ \delta \to 0 \ for \ N \ fixed.$$

For any fixed value of k (or bounded set of values of k)

$$\widehat{D_N G}(k) \to 1 \ as \ N \to \infty \ \ for \ \delta \ fixed \ and \ as \ \delta \to 0 \ \ for \ N \ fixed$$

3.4.1 Proof

This follows from a plot.

3.5 Discrete Differential Filters

We consider both the periodic case and the case of internal flow with no slip boundary conditions. In the periodic case, $\Omega = (0, L)^d, d = 2, 3$ and pressure and velocity spaces are, respectively,

$$Q = L_0^2(\Omega) := \{q \in L^2, \int_\Omega q = 0\},$$

$$X = H_\#^1(\Omega) := \{v \in H^1(\Omega) \cap L_0^2(\Omega) : \ v \ \text{is L periodic} \},$$

while in the case of internal flow Ω is a regular, bounded, polyhedral domain in $\mathbb{R}^d$ and

$$Q = L_0^2(\Omega),$$
$$X = H_0^1(\Omega) := \{v \in H^1(\Omega) : v|_{\partial\Omega} = 0\}.$$

The space of divergence free functions is denoted

$$V := \{v \in X, (\nabla \cdot v, q) = 0 \ \ \forall q \in Q\}.$$

The velocity-pressure finite element spaces $X^h \subset X$, $Q^h \subset Q$ are assumed to be conforming and satisfy the LBB^h condition, e.g. [G89]. The discretely divergence free subspace of X^h is, as usual

$$V^h = \{v^h \in X^h, (\nabla \cdot v^h, q^h) = 0 \ \ \forall q^h \in Q^h\}.$$

In addition, we assume the following approximation properties,[BS94]:

$$\inf_{v \in X_h} \|u - v\| \leq Ch^{k+1}|u|_{k+1}, \ \ u \in H^{k+1}(\Omega)^d,$$

$$\inf_{v \in X_h} \|u - v\|_1 \leq Ch^k|u|_{k+1}, \ \ u \in H^{k+1}(\Omega)^d,$$

$$\inf_{r \in Q_h} \|p - r\| \leq Ch^{s+1}|p|_{s+1}, \ \ p \in H^{s+1}(\Omega).$$

Taylor-Hood elements (see e.g. [BS94, G89]) are one common example of such a choice for (X^h, Q^h), and are also the elements we use in our numerical experiments.

We define next the discrete differential filter, following Manica and Kaya-Merdan [MM06]. Given $v \in L^2(\Omega)$, for a given filtering radius $\delta > 0$, $\overline{v}^h = A_h^{-1}v$ is the unique solution in X^h of

$$\delta^2(\nabla \overline{v}^h, \nabla \chi) + (\overline{v}^h, \chi) = (v, \chi) \ \ \forall \chi \in X^h. \tag{3.11}$$

Definition 25. Define the L^2 projection $\Pi^h : L^2(\Omega) \to X^h$ and discrete Laplacian operator $\Delta_h : X \to X^h$ in the usual way by

$$(\Pi^h v - v, \chi) = 0, \qquad (\Delta_h v, \chi) = -(\nabla v, \nabla \chi) \ \ \forall \chi \in X^h. \tag{3.12}$$

With Δ_h, we can write $\overline{v}^h = (-\alpha^2 \Delta_h + \Pi^h)^{-1}v$ and $A_h = (-\alpha^2 \Delta_h + \Pi^h)$. If Π^h is the L^2 projection on X^h and Π_1^h the projection on X^h with respect to the H^1 semi-inner product (grad , grad) then $\Delta_h = \Pi^h \Delta_h = \Delta_h \Pi_1^h$. Further, Δ_h is extended from X^h to X by zero on the orthogonal complement of X^h with respect to (grad , grad).

Remark 26. An attractive alternative is to define the differential filter by a discrete Stokes problem so as to preserve incompressibility approximately. In this case, given $\phi \in V$, $\overline{\phi}^h \in V^h$ would be defined by

$$\delta^2(\nabla\overline{\phi}^h, \nabla\chi) + (\overline{\phi}^h, \chi) = (\phi, \chi) \text{ for all } \chi \in V^h.$$

Herein we study (3.11) which seems to be perfectly acceptable when working with the rotational form of the nonlinearity.

We begin by recalling from [BIL06, MM06] some basic facts about discrete differential filters.

Lemma 27. *For $v \in X$, we have the following bounds for the discretely filtered and approximately deconvolved v*

$$\|\overline{v}^h\| \leq \|v\|, \quad \|\nabla\overline{v}^h\| \leq \|\nabla v\| \quad and \quad \|\nabla \times \overline{v}^h\| \leq \|\nabla v\|. \tag{3.13}$$

Proof. Set $\chi = \overline{v}^h$ and use standard inequalities. □

Finite element convergence theory can be extended to discrete differential filters. For example, we can prove the following.

Lemma 28. *For $\phi \in H_0^1(\Omega) \cap H^2(\Omega)$*

$$\delta^2\|\nabla(\phi - \overline{\phi}^h)\|^2 + \|\phi - \overline{\phi}^h\|^2$$

$$\leq C \inf_{v^h \in X^h} \{\delta^2\|\nabla(\phi - v^h)\|^2 + \|\phi - v^h\|^2\} + C\delta^4\|\triangle\phi\|^2.$$

Proof. The functions ϕ and $\overline{\phi}^h$ satisfy respectively for any $v^h \in X^h$

$$\delta^2(\nabla\overline{\phi}^h, \nabla v^h) + (\overline{\phi}^h, v^h) = (\phi, v^h),$$

$$\delta^2(\nabla\phi, \nabla v^h) + (\phi, v^h) = -\delta^2(\triangle\phi, v^h) + (\phi, v^h).$$

If the error is denoted $e := \phi - \overline{\phi}^h$, subtraction gives for any $v^h \in X^h$

$$\delta^2(\nabla e, \nabla v^h) + (e, v^h) = -\delta^2(\triangle\phi, v^h).$$

The rest of the proof follows standard FEM error analysis. Let $\widetilde{\phi} \in X^h$ be arbitrary and split the error as $e = (\phi - \widetilde{\phi}) - (\overline{\phi}^h - \widetilde{\phi})$. Rearranging the above error equation following that splitting, setting and using the Cauchy-Schwarz-Young inequality and the triangle inequality completes the proof. □

3.6   Reversibility of Approximate Deconvolution Models

"It remains to call attention to the chief remaining difficulty of our subject."
    H. Lamb, 1925.

The van Cittert approximate deconvolution operator satisfies

$$u = D_N \overline{u} + O(\delta^{2N+2}), \text{ for smooth } u \text{ and thus} \tag{3.14}$$

$$\overline{uu} \simeq \overline{D_N \overline{u} D_N \overline{u}} + O(\delta^{2N+2}) \tag{3.15}$$

in the smooth flow regions giving the family of Stolz-Adams-Kleiser deconvolution model for the closure problem. If R denotes the usual sub-filter stress tensor $R(u, u) := \overline{uu} - \overline{u}\overline{u}$, then this closure approximation is equivalent to the closure model

$$R(u, u) \Leftarrow R_N(\overline{u}, \overline{u}) := \overline{D_N \overline{u} D_N \overline{u}} - \overline{u}\overline{u}. \tag{3.16}$$

Definition 29. A tensor function $R(u, v)$ of two vector variables is reversible if

$$R(-u, -v) = R(u, v).$$

The interest in reversibility is that the true subgrid stress tensor τ is reversible. Thus, many feel that appropriate closure models should (at least to leading order effects) share the property. We next show that the ADM is both.

Lemma 30. *Consider the Stolz-Adams-Kleiser closure model in the case of periodic boundary conditions. For each $N = 0, 1, 2, \ldots$, the Stolz Adams-Kleiser closure model τ_N is reversible.*

Proof. Reversibility is immediate.   □

3.7   The Zeroth Order Model

The zeroth order closure ADM arises from the $N = 0$ operator $D = I$. It is

$$\nabla \cdot w = 0 \text{ and } w_t + \nabla \cdot (\overline{w\,w}) - \nu \Delta w + \nabla q = \overline{f}. \tag{3.17}$$

In some sense, it is the most basic (hence zeroth) model in LES. It also can arise by dropping the cross and Reynolds terms and keeping only the resolved term. It is the zeroth Stolz-Adams-Kleiser ADM model. It is the rational model truncated to $O(\delta^2)$ terms.

The theory of the whole family of ADMs begins, like the Leray theory of the Navier–Stokes equations, with a clear global energy balance. This is the key to the development of this family of models (or any model for that matter). It is clearest for the periodic problem. Proceeding formally, let (w, q) denote a periodic solution of the zeroth order model. Let the filter be $G = (-\delta^2 \triangle + 1)^{-1}$ and $A := (-\delta^2 \triangle + 1)$. Multiplying by Aw and integrating over the flow domain gives

$$\int_\Omega w_t \cdot Aw + \nabla \cdot (\overline{w}\,\overline{w}) \cdot Aw - \nu \triangle w \cdot Aw + \nabla q \cdot Aw \, dx = \int_\Omega \overline{f} \cdot Aw \, dx.$$

The nonlinear term exactly vanishes because

$$\int_\Omega \nabla \cdot (\overline{w}\,\overline{w}) \cdot Aw \, dx = \int_\Omega A^{-1}(\nabla \cdot (w\,w)) \cdot Aw \, dx$$

$$= \int_\Omega \nabla \cdot (w\,w) \cdot (A^{-1}Aw) \, dx$$

$$= \int_\Omega \nabla \cdot (w\,w) \cdot w \, dx$$

$$= \int_\Omega w \cdot \nabla w \cdot w \, dx = 0.$$

Integrating by parts the remaining terms gives

$$\frac{d}{dt}\{\|w(\cdot, t)\|_{L^2(\Omega)}^2 + \delta^2 \|\nabla w(\cdot, t)\|_{L^2(\Omega)}^2\}$$

$$+ \nu\{\|\nabla w(\cdot, t)\|_{L^2(\Omega)}^2 + \delta^2 \|\triangle w(\cdot, t)\|_{L^2(\Omega)}^2\} = \int_\Omega f \cdot w \, dx.$$

Integrating this gives an energy equality which has parallels to that of the NSE. In particular, we can clearly identify three physical quantities of model kinetic energy, energy dissipation rate and power input given, respectively, by

$$E_{model}(t) = \frac{1}{|\Omega|}\frac{1}{2}\int_\Omega |w(x, t)|^2 + \delta^2 |\nabla w(x, t)|^2 dx,$$

$$\epsilon_{model}(t) = \frac{1}{|\Omega|}\int_\Omega \nu|\nabla w(x, t)|^2 + \nu\delta^2 |\triangle w(x, t)|^2 dx,$$

$$P_{model}(t) = \frac{1}{|\Omega|}\int_\Omega f(x, t) \cdot w(x, t) dx.$$

This estimate, combined with a construction now standard for the Navier–Stokes equations, is strong enough to prove existence, uniqueness and regularity of strong solution of the zeroth order model, [LL06a]. To motivate the work that follows we first derive the following energy equality for strong solutions.

Proposition 31. *Let $u_0 \in L_0^2(\Omega)$, $f \in L^2(\Omega \times (0,T))$, and $\int_\Omega f(x,t)dx = 0$. For $\delta > 0$, let the averaging be $(-\delta^2 \triangle + 1)^{-1}$. Then, if w is a strong solution of the zeroth order model, w satisfies*

$$\frac{1}{2} \left[\|w(t)\|^2 + \delta^2 \|\nabla w(t)\|^2 \right] + \int_0^t \nu \|\nabla w(t')\|^2 + \nu\delta^2 \|\triangle w(t')\|^2 \, dt'$$

$$= \frac{1}{2} \left[\|\overline{u}_0(t)\|^2 + \delta^2 \|\nabla \overline{u}_0(t)\|^2 \right] + \int_0^t (f(t'), w(t'))dt'. \quad (3.18)$$

Proof. Motivated by the energy estimate, multiply the equation model by $(-\delta^2 \triangle + 1)w$, and integrate over the domain Ω. This gives

$$(w_t, (-\delta^2 \triangle + 1)w) + (\nabla \cdot (\overline{ww}), (-\delta^2 \triangle + 1)w)$$

$$+ (\nabla q, (-\delta^2 \triangle + 1)w) - (\nu \triangle w, (-\delta^2 \triangle + 1)w) = (\overline{f}, (-\delta^2 \triangle + 1)w). \quad (3.19)$$

The second term vanishes, as does the fourth term because $\nabla \cdot w = 0$ and the last term equals (f, w). Integrating by parts the first and third terms gives the differential equality

$$\frac{1}{2} \frac{d}{dt} \left[\|w(t)\|^2 + \delta^2 \|\nabla w(t)\|^2 \right] + \nu \|\nabla w(t)\|^2 + \nu\delta^2 \|\triangle w(t)\|^2 = (f(t), w(t)). \quad (3.20)$$

The results follows by integrating this from 0 to t.   □

Remark 32. The model thus has two terms which reflect extraction of energy from resolved scales. The energy dissipation in the model

$$\varepsilon_{model}(t) := \nu \|\nabla w(t)\|^2 + \nu\delta^2 \|\triangle w(t)\|^2 \quad (3.21)$$

is enhanced by the extra term $\nu\delta^2 \|\triangle w(t)\|^2$. This term dissipates energy locally where large curvatures in w occur, rather than large gradients. This term thus acts as an irreversible energy drain localized at large local fluctuations. The second term, $\delta^2 \|\nabla w(t)\|^2$, occurs in the models kinetic energy

$$E_{model}(t) := \frac{1}{2}\left[\|w(t)\|^2 + \delta^2 \|\nabla w(t)\|^2\right]. \qquad (3.22)$$

The true kinetic energy $\left(\frac{1}{2}\|w(t)\|^2\right)$ in regions of large deformations is thus extracted, conserved and stored in the kinetic energy penalty term $\delta^2 \|\nabla w(t)\|^2$. Thus, this reversible term acts as a kinetic "Energy sponge". Both terms have an obvious regularizing effect.

Remark 33. The key idea in the proof of the energy equality is worth noting and emphasizing. The kinetic energy in the Navier Stokes equations is bounded primarily because the nonlinear term $\nabla \cdot (uu)$ is a mixing term which redistributes kinetic energy rather than increasing it. Mathematically this is because of the skew symmetry property $(\nabla \cdot (uu), u) = 0$. The main idea in the proof is to lift this property of the Navier Stokes equations by deconvolution to understand the energy balance in the model. This is done as follows. Noting that all operations are self adjoint and, by definition of the averaging used, $(-\delta^2\triangle + 1)\overline{\phi} = \overline{(-\delta^2\triangle + 1)\phi} = \phi$. We have

$$(\nabla \cdot (\overline{ww}), (-\delta^2\triangle + 1)w) = (\nabla \cdot (ww), \overline{(-\delta^2\triangle + 1)w}) = (\nabla \cdot (ww), w) = 0. \qquad (3.23)$$

The stability bound in the Proposition is very strong. Using Galerkin approximations and this stability bound to extract a limit, it is straight forward to prove existence for the model (following [LL03]).

We first prove the following uniqueness result.

Theorem 34. *Assume that the data is regular. Then there exists at most one solution to the model.*

3.7.1 Proof

Let (w_1, q_1) and (w_2, q_2) be two solutions. Write $\phi = w_2 - w_1$, $\delta q = q_2 - q_1$. Thus ϕ is solution to the problem

$$\phi_t + \nabla \cdot (\overline{w_2 w_2 - w_1 w_1}) - \nu\triangle\phi + \nabla\delta q = 0, \qquad (3.24)$$

$$\nabla \cdot \phi = 0,$$

$$\phi_{t=0} = 0,$$

subject to periodic boundary conditions with zero mean. Notice that by using Schwartz rule in the absence of boundaries one has in the sense of the distributions

$$\nabla \cdot (\overline{w_2 w_2 - w_1 w_1}) = A^{-1} \nabla \cdot (w_2 w_2 - w_1 w_1).$$

Using $A\phi$ as test function and integrating yields

$$\frac{d}{2dt} \int (|\phi|^2 + \delta^2 |\nabla \phi|^2) + \nu \int (|\nabla \phi|^2 + \delta^2 |\Delta \phi|)^2$$

$$= - \int A^{-1} \nabla \cdot (w_2 w_2 - w_1 w_1) \cdot A\phi. \quad (3.25)$$

Using self-adjointness of A gives

$$\int A^{-1} \nabla \cdot (w_2 w_2 - w_1 w_1) \cdot A\phi = \int \nabla \cdot (w_2 w_2 - w_1 w_1) \cdot \phi.$$

Furthermore, using the incompressibility constraint, one obtains after an easy algebraic computation and an integration by parts,

$$\int \nabla \cdot (w_2 w_2 - w_1 w_1) \cdot \phi = - \int \phi \cdot \nabla \phi \cdot w_1.$$

Finally,

$$\frac{d}{2dt} \int (|\phi|^2 + \delta^2 |\nabla \phi|^2) + \nu \int (|\nabla \phi|^2 + \delta^2 |\Delta \phi|)^2 = - \int \phi \cdot \nabla \phi \cdot w_1. \quad (3.26)$$

By the Cauchy-Schwarz inequality,

$$\left| \int \phi \cdot \nabla \phi \cdot w_1 \right| \leq ||w_1||_{(L^4)^3} ||\phi||_{(L^4)^3} ||\nabla \phi||_{(L^2)^3}.$$

Since

$$w_1, w_2, \phi \in L^2([0,T], (H^2)^3) \cap L^\infty([0,T], (H^1)^3) \subset L^\infty([0,T], (L^4)^3),$$

(by using Sobolev embedding theorem) it follows that

$$||w_1||_{L^4} ||\phi||_{(L^4)^3} ||\nabla \phi||_{(L^2)3} \leq C ||\nabla \phi||^2_{(L^2)^3},$$

where C is a constant which only depends on the data f and u_0. Finally, with $C' = C'(\delta)$

$$C ||\nabla \phi||^2_{L^2} \leq C'^2 + \delta^2 ||\nabla \phi||^2.$$

By putting all of this together gives

$$\frac{d}{dt}\int (|\phi|^2 + \delta^2|\nabla\phi|^2) \leq C\int (|\phi|^2 + \delta^2|\nabla\phi|^2),$$

with $C = C(\delta)$. Since $|\phi|^2 + \delta^2|\nabla\phi|^2$ vanishes when $t = 0$, Gronwall's Lemma implies that it vanishes for almost every t. Hence, uniqueness follows and the theorem is proven.

Theorem 35. *Consider the zeroth order model under periodic boundary conditions. Let the averaging operator be given by $(-\delta^2\triangle + 1)^{-1}$, $\delta > 0$ be fixed and suppose $u_0 \in H(\Omega)$, $f \in L^2(\Omega \times (0,T))$, with $\int_\Omega f(x,t)dx = 0$. For any $u_0 \in (L^2(\Omega))^3$ with zero mean and*

$$\nabla \cdot u_0 = 0, \ f \in L^2(\Omega \times (0,T)), \ with \ \int_\Omega f(x,t)dx = 0.$$

the zeroth order model has a unique L- periodic weak solution

$$(w,q) \in [L^2([0,T],(H^2)^3) \cap L^\infty([0,T],(H^1)^3)] \times L^2([0,T] \times \Omega) \quad (3.27)$$

and the energy equality holds:

$$\frac{1}{2}(||w(t)||^2 + \delta^2||\nabla w(t)||^2) + \int_0^t \nu(||\nabla w(t)||^2 + \delta^2||\triangle w(t')||^2)dt' \quad (3.28)$$

$$= \frac{1}{2}(||\overline{u}_0||^2 + \delta^2||\nabla\overline{u}_0||^2) + \int_0^t (\overline{f},w)dt'.$$

That weak solution is also a unique strong solution. Suppose additionally, $u_0 \in V \cap H^{k-1}(\Omega)$ and $f \in L^\infty(0,T;H^k(\Omega))$ Then the model's solution satisfies

$$w \in L^2(0,T;H^{k+2}(\Omega)) \cap L^\infty(0,T;H^{k+1}(\Omega)),$$

$$q \in L^2(0,T;H^k(\Omega)).$$

If $u_0 \in C_\#^\infty(\Omega \times (0,T))$, $\nabla \cdot u_0 = 0$ and $f \in C_\#^\infty(\Omega \times (0,T))$, then that solution is smooth, $(w,q) \in [C^\infty(\Omega \times (0,T))]^4$.

Let $w = w(\delta)$ denote this solution. Then, there is a subsequence $\delta_j \to 0$ and a weak solution of the Navier–Stokes equations such that [4]

$$w(\delta_j) \to u, \ as \ \delta_j \to 0.$$

---

[4] The convergence is in the norms of the natural energy spaces for the Navier–Stokes equations.

Proof. The proof of existence is easy once the energy balance is identified (we omit the rest of the proof, see [LL06a] for proofs.). Indeed, let ψ_r be the orthogonal basis for $H(\Omega)$ of eigenfunctions of the Stokes' operator under periodic with zero mean boundary conditions. These are also the eigenfunctions of $(-\delta^2\triangle+1)$ in the same setting. Thus, let $-\triangle\psi_r = \lambda_r\psi_r$. Let $H_k(\Omega) := span\{\psi_r : r = 1,\ldots,k\}$. The Galerkin approximation $w_k \in H_k(\Omega)$ satisfies

$$(w_{k,t}, \psi) + (\nabla \cdot (\overline{ww}), \psi) - \nu(\triangle w, \psi) = (\overline{f}, \psi), \forall \psi \in H_k(\Omega). \qquad (3.29)$$

As usual, the Galerkin approximation (3.29) reduces to a system of ordinary differential equations for the undetermined coefficients $C_{k,r}(t)$. Existence will follow from an á priori bound on its solution. Since $w_k \in H_k(\Omega)$, it follows that $(-\delta^2\triangle+1)w_k = \sum_{r=1}^{k}(\delta^2\lambda_r+1)C_{k,r}\psi_r(x) \in H_k(\Omega)$. Thus it is permissible to set $\psi = (-\delta^2\triangle+1)w_k$ in (3.29). By exactly the same proof as for Proposition 31 we have

$$\frac{1}{2}[\|w_k(t)\|^2 + \delta^2\|\nabla w_k(t)\|^2] + \int_0^t \nu\|\nabla w_k(t')\|^2 + \nu\delta^2\|\triangle w_k(t')\|^2 \, dt' \qquad (3.30)$$

$$= \frac{1}{2}[\|\overline{u}_0(t)\|^2 + \delta^2\|\nabla\overline{u}_0(t)\|^2] + \int_0^t (f(t'), w_k(t'))dt'.$$

The Cauchy-Schwarz inequality then immediately implies

$$\|w_k\|_{L^\infty(0,T;H^1(\Omega))} \leq M_1 = M_1(f, u_0, \delta) < \infty,$$

$$\|w_k\|_{L^\infty(0,T;L^2(\Omega))} \leq M_2 = M_2(f, u_0) < \infty, and$$

$$\|w_k\|_{L^2(0,T;H^2(\Omega))} \leq M_3 = M_3(f, u_0, \delta, \nu) < \infty,$$

and the system of approximate ODEs thus has a unique solution.

From the above á priori bounds and using exactly the same approach as in the Navier Stokes case, following the beautiful and clear presentation of Galdi [Ga00], letting $k \to \infty$ we recover a limit, w, which is (using the stronger á priori bounds) a unique strong solution of the model satisfying the energy equality and having the regularity[5]

$$w \in L^\infty(0, T; H^1(\Omega)) \cap L^2(0, T; H^2(\Omega)) \qquad (3.31)$$

for $\delta > 0$ and $\nu > 0$.   □

---

[5]It is interesting to note that in the next regularity result the spacial regularity of $H^1(\Omega)$ and $H^2(\Omega)$ arise because the filter G is second order.

It is also important not to forget that the zeroth order model is a mathematical testing ground and not of sufficient accuracy for most practical computations. Thus, it is more important to find models with similar strong mathematical properties which are more accurate. The critical condition "more accurate" is presently evaluated two ways: analytical studies in smooth flow regions in experimental studies in turbulent regions.

Next we turn to the modeling error in the simple model. Let $\overline{\tau}$ denote the modeling consistency error tensor

$$\overline{\tau}(u,\,u) := \overline{\overline{u}\,\overline{u}} - \overline{u\,u}$$

Then it is straightforward to see that the true flow averages $\overline{u}$ satisfy

$$\overline{u}_t + \nabla \cdot (\overline{u}\,\overline{u}) + \nabla \overline{p} - \nu \Delta \overline{u} = \overline{f} + \nabla \cdot \overline{\tau}$$

and the error in the model $e = \overline{u} - w$ satisfies an equation driven only by the averaged consistency error $\nabla \cdot \overline{\tau}$:

$$e_t + \nabla \cdot (\overline{u}\,\overline{u} - w\,w) + \nabla(\overline{p} - q) - \nu \Delta e = \nabla \cdot \overline{\tau}.$$

Thus, $\|e\|$ being small depends upon two factors: a small consistency error, $\|\overline{\tau}\|$ small, and a strong enough stability property that $\|e\|$ is bounded by some norm of $\overline{\tau}$. If the stability constants in this bound are to be independent of δ, then (with the analytic tools available at this time) an extra condition ensuring global uniqueness of u is necessary. The other criteria is that $\|\overline{\tau}\|$ is small as $\delta \to 0$. We next give analytical bounds verifying that $\overline{\tau} \to 0$ (with rates) as $\delta \to 0$ for smooth enough solutions u of the NSE. To give bounds for $\|\overline{u} - w\|$ we need a strong enough regularity condition upon u to apply Gronwall's equality uniformly in δ. A sufficient condition for this is

$$\|\nabla u\| \in L^4(0, T). \tag{3.32}$$

This can obviously be weakened in many ways. Next we need a strong enough regularity condition on u to extract a bound on the models *consistency error evaluated at the true solution* in the norm $L^2(\Omega \times (0, T))$, i.e., on

$$\|uu - \overline{uu}\|_{L^2(\Omega \times (0,T))}. \tag{3.33}$$

It will turn out that a sufficient condition for this to be $O(\delta^2)$ is

$$u \in L^4(0, T; H^2(\Omega)). \tag{3.34}$$

To proceed, filtering the Navier Stokes' equations shows that $\overline{u}$ satisfies, after rearrangement,

$$\nabla \cdot \overline{u} = 0, \quad \text{and} \quad \overline{u}_t + \nabla \cdot (\overline{u}\overline{u}) + \nabla \overline{p} - \nu \triangle \overline{u} = \overline{f} - \nabla \cdot \overline{\tau_0} \tag{3.35}$$

where τ_0 is the consistency error term, in this case given by

$$\tau_0 := uu - \bar{u}\bar{u}. \tag{3.36}$$

Theorem 36. *Let the filtering be* $\bar{\phi} = (-\delta^2 \triangle + 1)^{-1}\phi$; *let u be a unique, smooth, strong solution of the Navier Stokes' equations. Then, there is a*

$$C^* = C^*(\nu, T, \|u\|_{L^4(0,T;H^2(\Omega))})$$

such that the modeling error $\phi := \bar{u} - w$ *satisfies*

$$\|\phi\|^2_{L^\infty(0,T;L^2(\Omega))} + \delta^2 \|\nabla\phi\|^2_{L^\infty(0,T;L^2(\Omega))} + \nu \|\nabla\phi\|^2_{L^2(0,T;L^2(\Omega))}$$
$$+ \nu\delta^2 \|\triangle\phi\|^2_{L^2(0,T;L^2(\Omega))} \leq C^* \|\tau_0\|^2_{L^2(\Omega\times(0,T))}. \tag{3.37}$$

The modeling consistency error satisfies

$$\|\tau_0\|^2_{L^2(\Omega\times(0,T))} \leq C\delta^4 \|u\|^2_{L^4(0,T;H^2(\Omega))}. \tag{3.38}$$

Proof. First we note that by the definition of τ_0 and the Sobolev inequality

$$\|\tau_0\|_{L^2(\Omega\times(0,T))} = \|\overline{uu} - \bar{u}u + \bar{u}u - uu\| \leq \tag{3.39}$$
$$\leq 2 \|u\|_{L^\infty(\Omega)} \|u - \bar{u}\| \leq 2\delta^2 \|\triangle u\|^2.$$

Thus, by squaring and integrating, we have, as claimed: $\|\tau_0\|^2_{L^2(\Omega\times(0,T))} \leq C\delta^4 \|u\|^2_{L^4(0,T;H^2(\Omega))}$.

For the proof that the modeling error is bounded by model consistency error τ_0, we subtract and mimic the proof of the models energy estimate. Indeed, subtracting equations for w from the above equation for $\bar{u}$ shows that the modeling error ϕ satisfies $\phi(0) = 0, \nabla \cdot \phi = 0$ and

$$\phi_t + \nabla \cdot (\overline{uu} - ww) + \nabla(\bar{p} - q) - \nu\triangle\phi = -\nabla \cdot \overline{\tau_0}, \text{ in } \Omega \times (0,T). \tag{3.40}$$

Under the above regularity assumptions, u is a strong solution of the Navier Stokes' equations and ϕ satisfies the above equation strongly also. Thus, only two paths are reasonable to bound ϕ by τ_0:

(a) Multiply by ϕ and integrate.
(b) Multiply by $(-\delta^2\triangle + 1)\phi$ and integrate.

Following the proof of the energy estimate we do (b). This gives, after steps which follow exactly those in the energy estimate,

$$\frac{1}{2}\frac{d}{dt}[\|\phi\|^2 + \delta^2 \|\phi\|^2] + \nu[\|\nabla\phi\|^2 + \|\triangle\phi\|^2] = \tag{3.41}$$

$$\int_\Omega (\overline{uu} - ww) : \nabla\phi \, dx + \int_\Omega \tau_0 : \nabla\phi \, dx. \tag{3.42}$$

The third term is handled in the standard way: adding and subtracting $w\overline{u}$. This gives

$$\int_{\Omega} (\overline{uu} - ww) : \nabla\phi \, dx = \int_{\Omega} \phi \cdot \nabla\phi \cdot \overline{u} \, dx. \tag{3.43}$$

Next, use the following inequalities, which are valid in two and three dimensions (and improvable in two dimensions),

$$\left| \int_{\Omega} \phi \cdot \nabla\phi \cdot \overline{u} \, dx \right| \leq \frac{\nu}{4} \|\nabla\phi\|^2 + C(\nu) \|\overline{u}\|^4 \|\phi\|^2, \tag{3.44}$$

$$\left| \int_{\Omega} \tau_0 : \nabla\phi \, dx \right| \leq \frac{\nu}{4} \|\tau_0\|^2 + C(\nu) \|\nabla\phi\|^2. \tag{3.45}$$

These give

$$\frac{1}{2}\frac{d}{dt}[\|\phi\|^2 + \delta^2 \|\phi\|^2] + \nu[\|\nabla\phi\|^2 + \|\triangle\phi\|^2]$$

$$\leq C(\nu) \|\tau_0\|^2 + C(\nu) \|\overline{u}\|^4 \|\phi\|^2. \tag{3.46}$$

The theorem then follows by Gronwall's inequality. □

3.8 Remarks

The results on the ADM family are remarkable because these are exactly properties which any reasonably derived sub-model for the Navier Stokes equations should satisfy but which either have not been proven or likely are not true for most models in large eddy simulation. Again, the key idea is a simple and direct connection between the models kinetic energy and the kinetic energy of the Navier Stokes equations through the deconvolution operator. The zeroth order model is a mathematical toy but it is also the mathematical test bed where the key ideas for the whole family of ADMs were developed since (as noted in the preface) the zeroth order model and a general ADM are related by a simple change of variable).

The originators of ADMs in LES are Adams, Kleiser and Stolz. Some of the important early papers include [SA99,SAK01a,SAK01b,AS01,AS02,SAK02]. Most of this book is concerned with LES models and regularizations based on the basic van Cittert deconvolution method and differential filters. The analysis of discrete differential filters began in [LMNR08, MM06, LMNR09, CL10]; there are still many mathematical gaps between the simplest case of periodic BCs with continuous differential filters and the real case of no slip BCs with incompressibility and discrete differential filters. There is certainly great opportunities for improving deconvolution models based on other deconvolution operators. Already, algorithmically simple modifications have produced great accuracy improvements (e.g., [LS07, LS09]). Existence theory for ADMs began in [LL03] and continued in [D04, LL06a, DE06, Le06, LP09] and [S08] (which currently contains the most general results). The results in this last paper provide an important theoretical guide in the search for better deconvolution operators for LES. For further developments in Tikhonov deconvolution and variants in LES see [MS09].

Chapter 4
Phenomenology of ADMs

"The ultimate goal, however, must be a rational theory of statistical hydrodynamics where.....properties of turbulent flows can be mathematically deduced from the fundamental equations of hydromechanics." - E. Hopf

4.1 Basic Properties of ADMs

An approximate deconvolution operator denoted by D is an approximate filter inverse that is accurate on the smooth velocity components and does not magnify the rough components.

Definition 37. The higher order generalized fluctuation is

$$w^* := w - D(\overline{w})$$

Given a chosen deconvolution operator D, the associated ADMs are given by

$$w_t + \nabla \cdot (\overline{D_N w \ D_N w}) - \nu \triangle w + \nabla q + \chi w^* = \overline{f}, \ and \ \nabla \cdot w = 0. \quad (4.1)$$

The w^* term is included to damp strongly the temporal growth of the fluctuating component of w driven by noise, numerical errors, and inexact boundary conditions. The two simplest examples of such a models arise:

(a) *when $N = 0$ and the w^* term is dropped.* This zeroth order model also arises as the zeroth order model in many different families of LES models. It is given by

$$w_t + \nabla \cdot (\overline{w \ w}) - \nu \triangle w + \nabla q = \overline{f}, \ and \ \nabla \cdot w = 0. \quad (4.2)$$

(b) *when no closure model is used on the nonlinear term and the time relaxation regularization is added*

W.J. Layton and L.G. Rebholz, *Approximate Deconvolution Models of Turbulence*, Lecture Notes in Mathematics 2042, DOI 10.1007/978-3-642-24409-4_4, © Springer-Verlag Berlin Heidelberg 2012

$$w_t + \nabla \cdot (w\, w) - \nu \triangle w + \nabla q + \chi w^* = f, \ and \ \nabla \cdot w = 0. \qquad (4.3)$$

The Stolz-Adams-Kleiser ADMs are highly accurate, in the sense that their consistency error is asymptotically small as δ approaches zero, they are reversible, conserve model energy and helicity, and can be shown to be approximately Galilean invariant to high order. Their usefulness thus hinges on their stability properties.

To see the mathematical key to the estimates of energy and helicity dissipation rates we first recall from the energy equality for the ADM [DE06]. Paralleling the case of the zeroth order model, the natural idea is to make the nonlinear term disappear in the energy inequality by multiplying the model by $AD_N w$ then integrating over the flow domain and integrating by parts. To give a hint of why this procedure is hopeful, we note that, in effect, this is simply re-norming the space $L^2(\Omega)$.

Recall the notation for the deconvolution weighted norm and inner product,

$$(\phi, \psi)_N := (\phi, D_N \psi), \quad \|\phi\|_N := (\phi, \phi)_N^{1/2} = (\phi, D_N \phi)^{1/2}.$$

Lemma 38. *The norm $\|v\|_{D_N}$ is a norm on $L^2(\Omega)$ which is equivalent to the $L^2(\Omega)$ norm.*

Proof. This follows from the fact that the transfer function of D_N, $\widehat{D_N}$, is bounded and bounded away from zero. $\qquad \square$

Proposition 39. *If w is a weak or strong solution[1], w satisfies*

$$\frac{1}{2}[\|w(T)\|_N^2 + \delta^2 \|\nabla w(T)\|_N^2] + \int_0^T \nu \|\nabla w(t)\|_N^2 + \nu \delta^2 \|\triangle w(t)\|_N^2 dt$$

$$= \frac{1}{2}[\|\overline{u}_0\|_N^2 + \delta^2 \|\nabla \overline{u}_0\|_N^2] + \int_0^T (f, w(t))_N \, dt.$$

Proof. (Sketch) Let (w, q) denote a periodic solution of the ADM (1.4). Multiplying by $AD_N w$ and integrating over Ω gives

---

[1] Unlike the NSE case, it is known that weak=strong for the ADM and both exist and are unique.

$$\int_\Omega (w_t \cdot AD_N w + \nabla \cdot (\overline{D_N w\, D_N w}) \cdot AD_N w - \nu \triangle w \cdot AD_N w + \nabla q \cdot AD_N w)\, dx$$

$$= \int_\Omega \overline{f} \cdot AD_N w\, dx.$$

The nonlinear term vanishes exactly because

$$\int_\Omega \nabla \cdot (\overline{D_N w\, D_N w}) \cdot AD_N w\, dx = \int_\Omega A^{-1}(\nabla \cdot (D_N w\, D_N w)) \cdot AD_N w\, dx$$

$$= \int_\Omega \nabla \cdot (D_N w\, D_N w) \cdot D_N w\, dx = 0.$$

Integrating by parts the remaining terms gives

$$\frac{d}{dt} \frac{1}{2} \{ \|w(t)\|_N^2 + \delta^2 \|\nabla w(t)\|_N^2 \} + \nu \{ \|\nabla w(t)\|_N^2 + \delta^2 \|\triangle w(t)\|_N^2 \} = (f, w(t))_N.$$

(4.4)

The result follows by integrating this from 0 to t. □

From Proposition 3.1, the ADMs kinetic energy, $E_{ADM-N}(w)(t)$, energy dissipation rate, $\varepsilon_{ADM-N}(w)(t)$, time averaged dissipation rate, $< \varepsilon_{ADM-N} >$, and power input, $P_{ADM-N}(w)(t)$, are clearly identified.

$$E_{ADM-N}(w)(t) := \frac{1}{2|\Omega|} \{ \|w(t)\|_N^2 + \delta^2 \|\nabla w(t)\|_N^2 \}, \qquad (4.5)$$

$$\varepsilon_{ADM-N}(w)(t) := \frac{\nu}{|\Omega|} \{ \|\nabla w(t)\|_N^2 + \delta^2 \|\triangle w(t)\|_N^2 \}, \qquad (4.6)$$

$$< \varepsilon_{ADM-N} > := < \varepsilon_{ADM-N}(w)(t) >, \qquad (4.7)$$

$$P_{ADM-N}(w)(t) := \frac{1}{|\Omega|} (f, w(t))_N. \qquad (4.8)$$

4.2 The ADM Energy Cascade

The first question is: Does the model satisfy the conditions, listed earlier for the Navier–Stokes equations, which make existence of an energy cascade likely? Since the model (viewed in the right light) has the same nonlinearity as the Navier–Stokes equations, the conditions remaining are that (a) the model satisfies an energy equality, in which the model's kinetic energy and

energy dissipation are readily discernible, and (b) in the absence of viscosity
and external forcing, the model's kinetic energy is exactly conserved.

The previous section has revealed the model's kinetic energy, energy
dissipation rate, and power input. Using that D_N is self adjoint and positive,
we can rewrite these quantities by

$$E_{model}(w) := \frac{1}{2L^3}\left\{\left\|D_N^{\frac{1}{2}}w(t)\right\|^2 + \delta^2\left\|\nabla D_N^{\frac{1}{2}}w(t)\right\|^2\right\},$$

$$\epsilon_{model}(w) := \frac{\nu}{L^3}\left\{\left\|\nabla D_N^{\frac{1}{2}}w(t)\right\|^2 + \delta^2\left\|\triangle D_N^{\frac{1}{2}}w(t)\right\|^2\right\}, \qquad (4.9)$$

$$P_{model}(w) := \frac{1}{L^3}\left(D_N^{\frac{1}{2}}f(t), D_N^{\frac{1}{2}}w(t)\right). \qquad (4.10)$$

Lemma 40. *As $\delta \to 0$,*

$$E_{model}(w) \to E(w),$$

$$\varepsilon_{model}(w) \to \varepsilon(w), \; and$$

$$P_{model}(w) \to P(w).$$

Proof. As $\delta \to 0$ all the δ^2 terms drop out in the definitions above. Further,
examining the definition of the deconvolution operator D_N, it is clear that
$D_N \to I$ as $\delta \to 0$. Thus the result follows easily.   □

The mathematical development of the model has shown that the two key
conditions (of (a) an energy equality, and (b) energy conservation in the
absence of viscosity and energy input) are satisfied! Thus, we are on firm
theoretical ground to develop a quantitative similarity theory of the model,
along the lines of the K41 theory of turbulence. The theory of the whole
family of models begins, like the Leray theory of the Navier–Stokes equations,
with a clear global energy balance. This is the key to the development of this
family of models (or any model for that matter). It is clearest for the periodic
problem and for the zeroth order model. But a parallel theory, which we now
summarize, holds for the entire family.

We begin by studying the case without time relaxation: $\chi = 0$. We shall
return to the time relaxation term later to understand its effect on the
dynamics.

To this end, using dimensional analysis we investigated the energy
spectrum, E_{model}, for the approximate deconvolution model by making the
assumption that the following variables are involved:

- E_{model}: energy spectrum of model with $[E(k)] = [length]^3[time]^{-2}$
- ε_{model}: time averaged energy dissipation rate of the model's solution with
  $[\varepsilon_{model}] = [length]^2[time]^{-3}$
- k: wave number with $[k] = [length]^{-1}$

– δ: averaging radius with $[\delta] = [length]$

Assuming $L \gg \delta$ we state three similarity hypotheses for statistically isotropic and homogeneous turbulence generated by ADMs.
First similarity hypothesis: E_{model} is determined by $\bar{\varepsilon}$, δ, and k.
Second similarity hypothesis: At wave-numbers k considerably smaller than δ^{-1}, E_{model} is determined by $\bar{\varepsilon}$, and k but does not depend on δ.

The two hypotheses are similar to Kolmogorov's hypotheses.
Defining the two dimensionless parameters

$$\Pi_1 = k\delta$$

and

$$\Pi_2 = \frac{\Delta}{\delta}$$

and making use of the Π theorem we find that in the general case

$$E_{model} = \bar{\varepsilon}^{2/3} k^{-5/3} G\pi_1, \pi_2,$$

where G is a dimensionless function of two dimensionless variables.
The simplest case[2] is when $f(k\delta) = \alpha \; model$. In this case we have

$$E_{model}(k) = \alpha \; model \varepsilon_{model}^{2/3} k^{-5/3}.$$

It is not surprising that, since the ADM is dimensionally consistent with the Navier–Stokes equations, dimensional analysis would reveal a similar energy cascade for the model's kinetic energy. However, interesting conclusions result from the differences arise due to the energy sponge mechanism in the model's kinetic energy. To see this, consider the simplest case in which $f(k\delta) = \alpha_{model}$. Recall the definition of the model's kinetic energy. Since D_N is uniformly (in δ) equivalent to the identity, we have

$$E_{model}(w) := < \frac{1}{2}\left(\left\| D_N^{\frac{1}{2}} w \right\|^2 + \delta^2 \left\| \nabla D_N^{\frac{1}{2}} w \right\|^2 \right) > \simeq$$
$$< \frac{1}{2}\left[\|w(t)\|^2 + \delta^2 \|\nabla w(t)\|^2 \right] > \left(= \alpha \; model \varepsilon_{model}^{2/3} k^{-5/3} \right).$$

Further, $E_{model}(w)(k) \simeq (1 + \delta^2 k^2)E(w)(k)$. Thus we have:

---

[2]We stress that this simplification may or may not be true. It must be tested in careful numerical experiments.

$$E(w)(k) \simeq \frac{\alpha_{\ model}\varepsilon_{model}^{2/3}k^{-5/3}}{1+\delta^2 k^2}.$$

(More generally we have $E(w)(k) \simeq f(\delta k)\varepsilon_{model}^{2/3}k^{-5/3}/(1+ \delta^2 k^2)$.)This equation gives precise information about how small scales are truncated by the ADM. Indeed, there are two wave number regions depending on which term in the denominator is dominant: 1 or $\delta^2 k^2$. The transition point is clearly the cutoff wave number $k = \frac{1}{\delta}$. We thus have:

$$E(w)(k) \simeq \alpha_{\ model}\varepsilon_{model}^{2/3}k^{-5/3}, \text{ for } k \leq \frac{1}{\delta},$$

$$E(w)(k) \simeq \alpha_{\ model}\varepsilon_{model}^{2/3}\delta^{-2}k^{-11/3} , \text{ for } k \geq \frac{1}{\delta}.$$

4.2.1 Another Approach to the ADM Energy Spectrum

Suppose instead we assume that through the inertial range

$$E_{model}(k) = C\varepsilon_{model}^a k^b.$$

Then, equating units of both sides we have, just like in the NSE case,

$$E_{model}(k) = C\varepsilon_{model}^{2/3}k^{-5/3}.$$

Since

$$E_{model}(k) = (1+\delta^2 k^2)E(w)(k),$$

we have, through the inertial range:

$$E(w)(k) \simeq \frac{\alpha_{\ model}\varepsilon_{model}^{2/3}k^{-5/3}}{1+\delta^2 k^2}.$$

The main open question in this plausible approach is why do we not have instead (at the first step)

$$E_{model}(k) = C\varepsilon_{model}^a k^b \delta^c.$$

This is the justification for the longer and much more precise derivation we have taken above.

4.2.2   The ADM Helicity Cascade

Recent work of Ditlevsen and Giuliani [DG01a] [DG01b] and of Q. Chen,
S. Chen and Eyink [CCE03] has delineated the existence of a joint and
coupled energy and helicity cascade through the inertial range. Defining
helicity and the mean helicity dissipation rate of the NSE by

$$H(u)(t) := (u(t), \nabla \times u(t)), \qquad (4.11)$$

$$\gamma(u)(t) := \nu(\nabla \times u(t), \nabla \times (\nabla \times u(t))), \qquad (4.12)$$

then to oversimplify, (see [DG01a, DG01b, CCE03] for the interesting details)
decomposition into Fourier series and dimensional analysis as in the K41
theory gives for the NSE

$$H(k) = C < \gamma >< \epsilon >^{-1/3} k^{-5/3}$$

in the inertial range.

In a further investigation of physical fidelity of ADM's, the work in
[LMNR08] studies the helicity cascade in the ADM and found a result
analogous to the energy cascade result of the ADM: up to a filtering radius
dependent length scale, the ADM helicity cascade matches that of the NSE,
and after which helicity scales are truncated. We refer the reader to [LMNR08]
for the details.

4.3   The ADM Micro-Scale

The fundamental goal of LES models is to truncate scales so as to:

(a) Fit a turbulent flow simulation onto a given mesh and within a given
    amount of computational resources.
(b) Extract the maximum information on the underlying turbulent flow from
    that given computation.

Thus the question of a model's micro-scale is absolutely fundamental to
assessing effectiveness of an LES model. To begin this assessment, the model's
Reynolds numbers with respect to the model's largest and smallest scales are

$$\text{Large scales:} \quad Re_{model-large} = \frac{UL}{\nu(1 + (\frac{\delta}{L})^2)}$$

$$\text{Small scales:} \quad Re_{model-small} = \frac{w_{small}\eta_{model}}{\nu\left(1 + \left(\frac{\delta}{\eta_{model}}\right)^2\right)}.$$

As in the Navier–Stokes equations, the ADMs energy cascade is halted by viscosity grinding down eddies exponentially fast when

Small scales $Re_{model-small} \simeq O(1)$, i.e., when

$$\frac{w_{small}\eta_{model}}{\nu\left(1+\left(\frac{\delta}{\eta_{model}}\right)^2\right)} \simeq 1.$$

This last equation allows us to solve for the characteristic velocity of the model's smallest persistent eddies w_{small} and eliminate it from subsequent equations. This gives $w_{small} \simeq \nu\left(1+\left(\frac{\delta}{\eta_{model}}\right)^2\right)/\eta_{model}$.

The second important equation determining the model's micro-scale comes from matching energy in to energy out. The rate of energy input to the largest scales is

$$\Pi_{model} = \frac{E_{model}}{\left(\frac{L}{U}\right)} = \frac{U^2\left(1+\left(\frac{\delta}{L}\right)^2\right)}{\left(\frac{L}{U}\right)} = \frac{U^3}{L}\left(1+\left(\frac{\delta}{L}\right)^2\right).$$

When the model reaches statistical equilibrium, the energy input to the largest scales must match the energy dissipation at the model's micro-scale which scales like $\varepsilon_{small} \simeq \nu\left(|\nabla w_{small}|^2 + \delta^2|\triangle w_{small}|^2\right) \simeq \nu\left(\frac{w_{small}}{\eta_{model}}\right)^2\left(1+\left(\frac{\delta}{\eta_{model}}\right)^2\right)$. Thus we have

$$\frac{U^3}{L}\left(1+\left(\frac{\delta}{L}\right)^2\right) \simeq \nu\left(\frac{w_{small}}{\eta_{model}}\right)^2\left(1+\left(\frac{\delta}{\eta_{model}}\right)^2\right).$$

Inserting the above formula for the micro-eddies characteristic velocity w_{small} gives

$$\frac{U^3}{L}\left(1+\left(\frac{\delta}{L}\right)^2\right) \simeq \frac{\nu^3}{\eta_{model}^4}\left(1+\left(\frac{\delta}{\eta_{model}}\right)^2\right)^3.$$

First note that the expected case in LES is when $\left(\frac{\delta}{L}\right)^2 \ll 1$ (otherwise the procedure should be considered a VLES).[3] In this case the LHS simplifies to just $\frac{U^3}{L}$. Next, with this simplification, the solution to this equation depends on which term in the numerator of the RHS is dominant: 1 or $\left(\frac{\delta}{\eta_{model}}\right)^2$. The

---

[3] Very Large Eddy Simulation. The estimates of the micro-scale are easily extended to this case too.

former case occurs when the averaging radius δ is so small that the model is very close to the Navier–Stokes equations so the latter is the expected case. In this case we have $\eta_{model} \simeq Re^{-\frac{3}{4}} L$, when $\delta < \eta_{model}$. In the expected case, solving for the micro-scale gives

$$\eta_{model} \simeq Re^{-\frac{3}{10}} L^{\frac{4}{10}} \delta^{\frac{3}{5}} \ , \text{ when } \delta > \eta_{model}.$$

4.3.1   Design of an Experimental Test of the Model's Energy Cascade

One question not resolved in the similarity theory pertains to the unknown, non-dimensional function $f(k\delta)$. The principle of economy of explanation suggests that $f(k\delta)$ is constant, as discussed above. On the other hand, we know of no clear mathematical reason why $f(k\delta)$ should be constant. This question can be resolved by numerical experiments on the model itself (not on the NSE) to establish the curve between the Π 's. Having this curve gives complete quantitative information. Suppose that the E_{model} is desired for conditions k_a and δ_a. The dimensionless group $(\Pi_2)_a$ can be immediately evaluated as $k_a\delta_a$. Corresponding to this value of $(\Pi_2)_a$, the value of $(\Pi_1)_a$ is read off the plot and $(E_{model})_a$ is then computed as $(\Pi_1)_a \varepsilon_a^{2/3} k_a^{-5/3}$.

4.4   Remarks

The analysis in this chapter presupposes two things. First, the relaxation term in the original model is zero. The intent of this term is clearly to further truncate the model's energy cascade. In the next section we study its precise effect on the model's energy cascade. Second, our conclusions are developed mostly for the case: $f(k\delta) = \alpha_{model}$. This is conjectured by the principle of parsimony. The analysis and conclusions are easily adapted to any form of $f(k\delta)$. Thus, it is an open problem to determine through careful numerical tests the correct form of $f(k\delta)$.

The work in this chapter is based on [LN06b]; see [LMNR08] for a more comprehensive analysis including the helicity cascade. Application of turbulence phenomenology to LES models originated in [Mus96]. Some rigorous mathematical connections between LES model phenomenology and their energy and helicity balances have been made in [LRS10].

Chapter 5
Time Relaxation Truncates Scales

5.1 Time Relaxation

"In setting up a finite difference grid or a finite wave number space, a turbulent threshold is in effect defined and the question is:

How do the equations know how to communicate with the molecular dissipation range?

One of course finds empirically that, without any provision for dissipation, the cascade of energy to the higher wave numbers ultimately increases the energy of the smallest wave resolvable by the grid. This energy has no place further to go, and ultimately the calculation departs from nature sufficiently to give intolerable truncation error."

J. Smagorinsky, page 119 in: Frontiers of Numerical Mathematics (ed: R.E. Langer, U Wisconsin press, 1960).

The fundamental requirement for a successful turbulent flow simulation is to truncate scales to those representable on a computationally feasible mesh without substantially changing the large flow structures. Thus *success* for a turbulence minimally requires that

$$\triangle x = \delta = \eta_{model}.$$

By this simple, minimal and reasonable standard, almost every turbulence model is an utter failure! Indeed, for the base ADM, the microscale of the basic ADM is insufficient for numerical simulations. We therefore turn to an added time relaxation term. Time relaxation is another clever idea of Stolz, Adams and Kleiser. The idea of time relaxation is that since deconvolved averages are already computed, they can be used to introduce model diffusion exactly at the marginally resolved scales so as to force $\triangle x = \delta = \eta_{model}$ and have marginal effect on the large scales. The ADM time relaxation term, added to the base ADM, can also be added just to the NSE or to any other turbulence model. The ADM time relaxation operator precisely truncates

W.J. Layton and L.G. Rebholz, *Approximate Deconvolution Models of Turbulence*, Lecture Notes in Mathematics 2042, DOI 10.1007/978-3-642-24409-4_5, © Springer-Verlag Berlin Heidelberg 2012

small solution scales without altering appreciably the solution's large scales. Because this term is not linked to closure of the nonlinear terms and is widely useful, we isolate its effects by studying the sizes of the persistent scales in the Navier–Stokes equations and time relaxation term.

Adding it to the NSE gives the time relaxation regularization:

$$w_t + w \cdot \nabla w + \nabla q + \nu \triangle w + \chi(w - D_N \overline{w}) = f, \quad \text{in } \Omega \times (0, T).$$

The term $w - D_N \overline{w}$ is a generalization of a simple fluctuation ($w' := w - \overline{w}$) term. The intent of including it is to force the fluctuations below $O(\delta)$ to be driven to zero exponentially fast as $t \to \infty$ without affecting the formal order of accuracy[1] of the model as an approximation of the underlying fluids model. The simplest interesting case is $N = 0$. Here $G_0 \overline{u} = \overline{u}$ represents the part of the velocity that can be represented on a mesh of meshwidth $O(\delta)$, i.e., the part varying over length scales $l \geq O(\delta)$, while $u' := u - \overline{u}$ represents the part of the velocity varying over scales $l \leq O(\delta)$. When $N = 0$ the above model reduces to

$$w_t + \overline{w \cdot \nabla w} + \nu \triangle w + \nabla q + \chi w' = f, \text{in } \Omega \times (0, T) .$$

By the definition of the differential filter, when $N = 0$, $u' = u - \overline{u} = -\delta^2 \triangle \overline{u}$ so the term $\chi w' = -\chi \delta^2 \triangle \overline{u}$ represents an added and smoothed model diffusion term.

It is not hard to discretize linear time relaxation. For example, suppressing spacial discretizations, one can time step by

$$\frac{w^{n+1} - w^n}{\triangle t} + w^n \cdot \nabla w^{n+1} - \nu \triangle w^{n+1} +$$

$$+ \nabla p^{n+1} + \chi(w^{n+1} - \overline{w}^n) = f^{n+1} \tag{5.1}$$

$$\text{and } \nabla \cdot w^{n+1} = 0. \tag{5.2}$$

Note that at each step the filter acts on a velocity known explicitly from previous time steps. This property is important for efficiency. This method was studied by Anitescu, Layton and Pahlevani [ALP04]. Even though it treats the filtered term explicitly, the overall method is unconditionally stable.

Proposition 41. *Suppose* $||\overline{v}|| \leq ||v||$, *for all square integrable* v. *Then the above method is unconditionally stable:*

---

[1]Formal order of accuracy here is interpreted in the usual sense of the consistency error in smooth regions as $\delta \to 0$. It does not mean details of solutions must match. Proving this involves understanding both what is predictable and how it is to be predicted.

$$\frac{1 + \chi \triangle t}{2} ||w^N||^2 + \triangle t \sum_{n=0}^{N-1} \frac{\nu}{2} ||\nabla w^{n+1}||^2$$

$$\leq \frac{1 + \chi \triangle t}{2} ||w^0||^2 + \triangle t \sum_{n=0}^{N-1} \frac{1}{2\nu} ||f^{n+1}||^2_{-1}.$$

Proof. Take the inner product with w^{n+1}. We have

$$(w^n \cdot \nabla w^{n+1}, w^{n+1}) = 0, (\nabla p^{n+1}, w^{n+1}) = -(p^{n+1}, \nabla \cdot w^{n+1}) = 0.$$

Thus

$$\frac{||w^{n+1}||^2 - (w^n, w^{n+1})}{\triangle t} + \nu ||\nabla w^{n+1}||^2$$

$$+ \chi(||w^{n+1}||^2 - (\overline{w}^n, w^{n+1})) = (f^{n+1}, w^{n+1}).$$

Applying the CS inequality and the arithmetic-geometric mean inequality on the three inner product terms gives

$$(f^{n+1}, w^{n+1}) \leq \frac{\nu}{2} ||\nabla w^{n+1}||^2 + \frac{1}{2\nu} ||f^{n+1}||^2_{-1},$$

$$(w^n, w^{n+1}) \leq \frac{1}{2} ||w^{n+1}||^2 + \frac{1}{2} ||w^n||^2,$$

$$(\overline{w}^n, w^{n+1}) \leq \frac{1}{2} ||w^{n+1}||^2 + \frac{1}{2} ||\overline{w}^n||^2 \leq \frac{1}{2} ||w^{n+1}||^2 + \frac{1}{2} ||w^n||^2.$$

This gives

$$\frac{1 + \chi \triangle t}{2} ||w^{n+1}||^2 - \frac{1 + \chi \triangle t}{2} ||w^n||^2 + \triangle t \frac{\nu}{2} ||\nabla w^{n+1}||^2 \leq \triangle t \frac{1}{2\nu} ||f^{n+1}||^2_{-1}.$$

Summing proves stability.   □

Since the NSE nonlinearity continuously pumps energy into those small scales, this intent can be realized by picking the time relaxation parameter to match time relaxation dissipation in the model to the rate of energy flow from the marginally resolved scales to the unresolved scales in the NSE. This chapter explores this simple and powerful addition. First we show that it truncates scales. The microscale of the above model obviously depends on χ. We shall show that, computing the microscale and setting $\delta = \eta_{model}$, gives a formula that determines the one parameter χ

$$\chi_{optimal} \simeq \frac{U}{L^{\frac{1}{3}}} 2^{N+1} \delta^{-\frac{2}{3}}.$$

We then turn to nonlinear variants on the above idea. These have great potential but there are also many important open questions which are noted in the section as they occur. The material in this chapter is based on work in [LN06a].

5.2   The Microscale of Linear Time Relaxation

"One cannot help get the feeling that perhaps turbulence itself could be studied digitally by experimenting with the threshold of turbulence"
    J. Smagorinsky , page 119 in: Frontiers of Numerical Mathematics (ed: R.E. Langer, U Wisconsin press, 1960).

In the study of the simplest case of Navier–Stokes equations+Linear Time Relaxation term, we shall suppose that *the Reynolds number is high enough that all dissipation and scale truncation is created by precisely this relaxation term* (up to negligible effects). In other words, the Kolmogorov micro-scale for the Navier–Stokes equations is larger than the micro-scale induced by the relaxation term. The linear time relaxation regularization model we consider is to find a L-periodic (in $\Omega = (0, L)^3$ and with zero mean) velocity and pressure satisfying

$$w(x, 0) = u_0(x), \text{ in } \Omega ,$$

$$\nabla \cdot w = 0, \text{in } \Omega \times (0, T) ,$$

$$w_t + w \cdot \nabla w + \nabla q + \nu \triangle w + \chi(w - D_N \overline{w}) = f, \text{in } \Omega \times (0, T) .$$

The relaxation coefficient χ must be specified and has units $\frac{1}{time}$.

Definition 42. Define $Hw := D(\overline{w})$. Given a fluid velocity w its fluctuation is $w' := w - \overline{w}$. Given a deconvolution operator D the generalized fluctuation $w - D(\overline{w})$ is denoted
$$w^* := w - D(\overline{w}).$$

To develop the phenomenology of the time relaxation regularization, the first question is: *Does the Navier–Stokes equations + time relaxation combination satisfy the conditions which make existence of an energy cascade likely?* Since the model has the same nonlinearity as the Navier–Stokes equations, the conditions remaining are that:

- The solution satisfies an energy equality, in which its kinetic energy and energy dissipation are readily discernible.
- The energy dissipation induced by time relaxation acts predominantly on small scales and is negligible on large ones.

- In the absence of *both viscosity and relaxation* (for $\nu = \chi = 0$) the model's kinetic energy is exactly conserved (analogous to NSE case, in the absence of viscosity).

These conditions are satisfied; thus, we proceed to develop a quantitative similarity theory of the model, along the lines of the K41 theory of turbulence. The derivation of these results involves the classical dimensional analysis arguments of Kolmogorov coupled with precise mathematical knowledge of the model's kinetic energy balance.

Lemma 43. *The bounded linear operator $H_N = D_N G$ and $I - H_N :=I - D_N G$ are both symmetric, positive semi-definite operators on $L_0^2(\Omega)$. For $w \in L_0^2(\Omega)$*

$$\int_\Omega (w - H_N w) \cdot w \; dx \geq 0, \quad and \quad \int_\Omega H_N w \cdot w \; dx \geq 0.$$

Proof. Since we are in the periodic case, the proof is done by a direct calculation using Fourier series. All the operators involved are functions of the operator $A = -\delta^2 \triangle + 1$ and thus commute. To begin, expand $w(x, t) = \sum_{\mathbf{k}} \widehat{w}(\mathbf{k}, t) e^{-i \mathbf{k} \cdot \mathbf{x}}$, where $\mathbf{k} = \frac{2\pi \mathbf{n}}{L}$ is the wave number and $\mathbf{n} \in \mathbb{Z}^3$. Then, by direct calculation using Parseval's equality

$$\frac{1}{2L^3} \int_\Omega (D_N G w) \cdot w \; dx = \frac{2\pi}{L} \sum_k \widehat{H}_N(k) E(w)(k) \; , \text{ where}$$

$$\widehat{H}_N(k) = \frac{1}{1 + z^2} \sum_{n=0}^{N} \left(1 - \frac{1}{1 + z^2}\right)^n \; , \text{ where } z = \delta k.$$

The expression for $\widehat{H}_N(k)$ can be simplified by summing the geometric series. This gives

$$\widehat{H}_N(k) = 1 - \left(\frac{z^2}{1 + z^2}\right)^{N+1} \; , \text{ where } z = \delta k \; .$$

Since z is real, $0 \leq \frac{z^2}{1+z^2} \leq 1$, and $0 \leq 1 - (\frac{z^2}{1+z^2})^{N+1} \leq 1$. Thus we have shown

$$0 \leq \int_\Omega (D_N G w) \cdot w dx \leq \int_\Omega |w|^2 dx.$$

Similarly, we show $0 \leq 1 - \widehat{H}_N(k) \leq 1$ and

$$0 \leq \int_\Omega (w - D_N G w) \cdot w dx \leq \int_\Omega |w|^2 dx,$$

which completes the proof. □

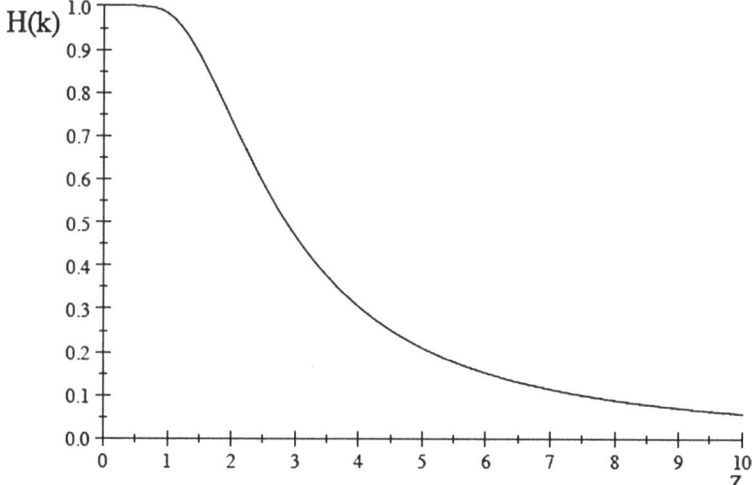

Fig. 5.1 Transfer function for $N = 5$

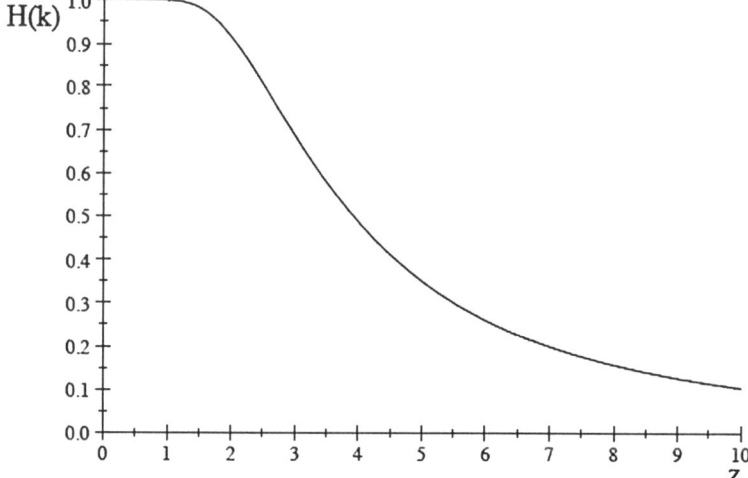

Fig. 5.2 Transfer function for $N = 10$

It is insightful to plot the transfer function

$$\widehat{H}_N(k) = 1 - \left(\frac{z^2}{1 + z^2} \right)^{N+1}$$

for a few values of N, which is done for $N = 5$, 10, and 100 in Figs. 5.1–5.3.

Examining these graphs, we observe the following (which can all easily be proven as well):

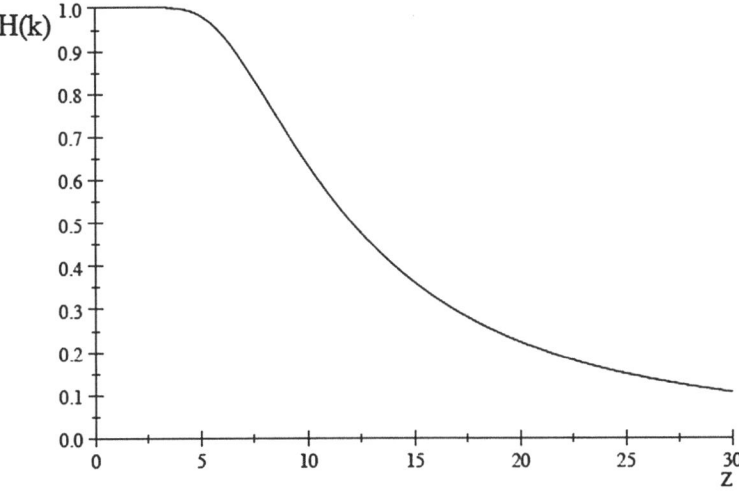

Fig. 5.3 Transfer function for $N = 100$

- $H_N w$ is very close to w for the low frequencies/largest solution scales.
- $H_N w$ attenuates small scales/high frequencies.
- The cutoff frequency (at which $\widehat{H}_N(k) \simeq 1/2$) grows slowly as $N \to \infty$.
- H_N is a compact operator since $\widehat{H}_N(k) \to 0$ as $k \to \infty$.

The theory of the time relaxation regularization begins with a clear global energy balance;

Proposition 44. *Let $u_0 \in L_0^2(\Omega)$, $f \in L^2(\Omega \times (0, T))$, and $\int_\Omega f(x, t)dx = 0$ for $\delta > 0$ and let the averaging be $(-\delta^2 \triangle + 1)^{-1}$. Then, if w is a strong solution of the time relaxation model, w satisfies*

$$\frac{1}{2}\|w(t)\|^2 + \int_0^t \int_\Omega \nu|\nabla w|^2 + \chi(w - D_N Gw) \cdot w \, dxdt' =$$

$$= \frac{1}{2}\|u_0\|^2 + \int_0^t \int_\Omega f \cdot w \, dxdt'.$$

Proof. Multiply the equation by w, integrate over the domain Ω, then integrate from 0 to t. $\square$

By the above lemma and energy estimate, since $(w - D_N Gw, w) > 0$ the model's relaxation term extracts energy. Thus, when Re is so large that viscous dissipation is negligible we define an induced energy dissipation rate for the model as

$$\varepsilon_{model}(w)(t) := \frac{1}{|\Omega|} \int_{\Omega} \chi(w - D_N Gw) \cdot w dx, \qquad (5.3)$$

$$\varepsilon_{model} := < \varepsilon_{model}(w)(t) > . \qquad (5.4)$$

The model's kinetic energy is the same as for the Navier–Stokes equations

$$E_{model}(w)(t) := \frac{1}{2} \|w(t)\|^2 . \qquad (5.5)$$

It is important to formulate the model phenomenology in a way that is as clear and correct as possible. To this end, the first step is to find the equivalent of the Reynolds number for the model. Recall the Reynolds number for the Navier Stokes equations is, in simplest terms, the ratio of nonlinearity to viscous terms evaluated at the largest solution scales (large velocity scale $= U$ and large length scale $= L$):

$$\text{for the NSE: } Re \simeq \frac{|u \cdot \nabla u|}{|\nu \triangle u|} \simeq \frac{U \frac{1}{L} U}{\nu \frac{1}{L^2} U} = \frac{UL}{\nu}.$$

The NSE's Reynolds numbers at the smallest persistent scales is obtained by replacing the large scales velocity and length by their small scales equivalent (small velocity scale $= u_{small}$ and small length scale $=$ microscale $= \eta$)

$$Re_{small} = \frac{u_{small}\eta}{\nu}.$$

To proceed we must find the physically appropriate and mathematically analogous quantity for the Navier–Stokes equations + time relaxation. The ratio of nonlinearity to dissipative effects should be the analogous quantity to Re provided dissipation due to time relaxation dominates molecular diffusion. Thus the Reynolds number for the NSE corresponds to

$$R_N \simeq \frac{|w \cdot \nabla w|}{|\chi(w - D_N Gw)|} : \text{evaluated at the largest scales.}$$

For example, if $N = 0$, and keeping in mind that for the large scales $(\frac{\delta}{L})^2 << 1$, then we calculate (using $w - \overline{w} = -\delta^2 \triangle \overline{w}$)

$$R_0 \simeq \frac{|w \cdot \nabla w|}{|\chi(w - \overline{w})|} = \frac{|w \cdot \nabla w|}{|\chi\delta^2 \triangle \overline{w}|}$$

$$= \frac{|w \cdot \nabla w|}{|\chi\delta^2 \triangle(-\delta^2 \triangle + 1)^{-1} w|} \simeq$$

$$\simeq \frac{U\frac{1}{L}U}{\chi\frac{\delta^2}{L^2}(\frac{\delta^2}{L^2}+1)^{-1}U}$$

$$= \frac{LU}{\chi\delta^2}\left(\frac{\delta^2}{L^2}+1\right) \simeq \frac{LU}{\chi\delta^2} = \frac{U}{\chi L}\left(\frac{\delta^2}{L^2}\right)^{-1}.$$

Thus for $N = 0$ we define

$$R_0 := \frac{U}{\chi L}\left(\frac{\delta^2}{L^2}\right)^{-1}.$$

In the general case, we have similarly that

$$R_N \simeq \frac{|w\cdot\nabla w|}{|\chi(w-D_NGw)|} \simeq \frac{U^2\frac{1}{L}}{\chi(\frac{\delta^2}{L^2})^{N+1}(\frac{\delta^2}{L^2}+1)^{-(N+1)}U}$$

$$= \frac{U}{\chi L}\left(\frac{\delta^2}{L^2}\right)^{-(N+1)}\left(\frac{\delta^2}{L^2}+1\right)^{N+1} \simeq \frac{U}{\chi L}\left(\frac{\delta^2}{L^2}\right)^{-(N+1)}.$$

Definition 45. The non-dimensionalized time relaxation parameter for the Navier–Stokes equations + time relaxation is

$$R_N = \frac{U}{\chi L}\left(\frac{\delta^2}{L^2}\right)^{-(N+1)} \quad, \text{ for } N = 0, 1, 2, \ldots.$$

Next we must form the small scales parameters which measure the ratio of nonlinearity to dissipation at the smallest persistent scales. Let w_{small} denote a characteristic velocity of the smallest persistent eddies and let η_{model} denote the length scale associated with them. Then, exactly as above we calculate

$$R_{N-small} \simeq \frac{|w_{small}\cdot\nabla w_{small}|}{|\chi(w_{small}-D_NGw_{small})|}$$

$$\simeq \frac{w_{small}^2\frac{1}{\eta_{model}}}{\chi\delta^{2N+2}(\frac{\delta^2}{\eta_{model}^2})^{N+1}(\frac{\delta^2}{\eta_{model}^2}+1)^{-(N+1)}w_{small}}$$

$$= \frac{w_{small}}{\chi\eta_{model}}\left(\frac{\delta^2}{\eta_{model}^2}\right)^{-(N+1)}\left(\frac{\delta^2}{\eta_{model}^2}+1\right)^{N+1}.$$

For the small scales it is no longer reasonable to suppose δ is small with respect to η_{model} since our goal is to force $\delta = \eta$. Thus the term $(\delta^2/\eta_{model}^2+1)^{N+1}$ is no longer negligible.

Definition 46. Let η_{model}, w_{small} denote, respectively, a characteristic length and velocity of the smallest persistent structures in the flow. The non-dimensionalized parameter associated with the smallest persistent scales of the Navier–Stokes equations + time relaxation is

$$R_{N-small} = \frac{w_{small}}{\chi \eta_{model}} \left(\frac{\delta^2}{\eta_{model}^2} \right)^{-(N+1)} \left(1 + \frac{\delta^2}{\eta_{model}^2} \right)^{N+1}.$$

Under K41 phenomenology, the energy cascade of the solution of the NSE + Time Relaxation is halted by dissipation caused by the time relaxation effects grinding down eddies exponentially fast when $R_{N-small} = O(1)$ at length-scale η_{model}. The estimate of the smallest resolved scales is based upon two principles:

- $R_{N-small} = O(1)$ at length-scale η_{model}.
- Statistical equilibrium: *rate of energy input at large scales = rate of energy dissipation at small scales.*

As in the Navier–Stokes case the power input at the large scales is like U^3/L. The dissipation at the smallest resolved scales is

$$\text{small scale dissipation rate} \simeq \chi(w - D_N Gw)w \simeq$$

$$\simeq \chi \delta^{2N+2}(\triangle^{N+1} A^{-(N+1)}w)w \simeq$$

$$\simeq \chi \left(\frac{\delta^2}{\eta_{model}^2} \right)^{N+1} \left(1 + \frac{\delta^2}{\eta_{model}^2} \right)^{-(N+1)} w_{small}^2$$

These two conditions thus give the pair of equations

$$\frac{w_{small}}{\chi \eta_{model}} \left(\frac{\delta^2}{\eta_{model}^2} \right)^{-(N+1)} \left(1 + \frac{\delta^2}{\eta_{model}^2} \right)^{N+1} \simeq 1$$

and

$$\frac{U^3}{L} \simeq \chi \left(\frac{\delta^2}{\eta_{model}^2} \right)^{N+1} \left(1 + \frac{\delta^2}{\eta_{model}^2} \right)^{-(N+1)} w_{small}^2 .$$

The first equation gives an estimate of the characteristic velocity of the smallest eddy in terms of the other parameters; solving for w_{small} gives

$$w_{small} \simeq \chi \eta_{model} \left(\frac{\delta^2}{\eta_{model}^2} \right)^{N+1} \left(1 + \frac{\delta^2}{\eta_{model}^2} \right)^{-(N+1)}.$$

Inserting this value into the second equation gives the following equation determining the model's micro-scale.

$$\frac{U^3}{L} \simeq$$

$$\simeq \chi \left(\frac{\frac{\delta^2}{\eta^2_{model}}}{1 + \frac{\delta^2}{\eta^2_{model}}} \right)^{N+1} \left[\chi \eta_{model} \left(\frac{\delta^2}{\eta^2_{model}} \right)^{N+1} \left(1 + \frac{\delta^2}{\eta^2_{model}} \right)^{-(N+1)} \right]^2$$

$$= \chi^3 \eta^2_{model} \left(\frac{\delta^2}{\eta^2_{model}} \right)^{3N+3} \left(1 + \frac{\delta^2}{\eta^2_{model}} \right)^{-(3N+3)} .$$

This is the fundamental equation determining the model's micro-scale. There are three cases: $\delta < \eta_{model}$, $\delta > \eta_{model}$ and $\delta = \eta_{model}$.

5.2.1   Case 1: Fully Resolved

In this case $\delta < \eta_{model}$ so that $1 + \delta^2/\eta^2_{model} \simeq 1$. The equation for the micro-scale reduces to

$$\frac{U^3}{L} \simeq \chi^3 \eta^2_{model} \left(\frac{\delta^2}{\eta^2_{model}} \right)^{3N+3} .$$

which implies

$$\eta_{model} \simeq \left(\frac{\chi^3 L}{U^3} \right)^{\frac{1}{6N+4}} \delta^{1 + \frac{1}{3N+2}} . \tag{5.6}$$

5.2.2   Case 2: Under Resolved

In this case $\delta > \eta$ so that $1 + \delta^2/\eta^2_{model} \simeq \delta^2/\eta^2_{model}$. We have then

$$\frac{U^3}{L} \simeq \chi^3 \eta^2_{model} \left(\frac{\delta^2}{\eta^2_{model}} \right)^{3N+3} \left(\frac{\delta^2}{\eta^2_{model}} \right)^{-(3N+3)} = \chi^3 \eta^2_{model}$$

$$\frac{U^3}{L} \simeq \chi \delta^{2N+2} \left(\frac{1}{\eta^2_{model}} \right)^{N+1} \left(\frac{\delta^2}{\eta^2_{model}} \right)^{N+1} \left[\frac{\chi \delta^{2N+2}}{\eta^N_{model} (\frac{\delta^2}{\eta^2_{model}})^{N+1}} \right]^2 ,$$

which gives, after simplification, for $N \geq 1$

$$\eta_{model} \simeq \sqrt{\frac{\chi^3 L}{U^3}} .$$

5.2.3   Case 3: Perfect Resolution

In this case $\delta = \eta_{model}$ so that $1 + \frac{\delta^2}{\eta_{model}^2} \simeq 2$. The interesting question is to determine the choice of relaxation parameter that enforces this case. Setting $\delta = \eta_{model}$ and solving for χ gives

$$\frac{U^3}{L} \simeq \chi^3 \eta_{model}^2 2^{-(3N+3)} = \chi^3 \delta^2 2^{-(3N+3)}.$$

Solving for χ gives

$$\chi_{optimal} \simeq \frac{U}{L^{\frac{1}{3}}} 2^{N+1} \delta^{-\frac{2}{3}}. \qquad (5.7)$$

In this case the consistency error of the relaxation term (evaluated for smooth flow fields) is, for this scaling of relaxation parameter,

$$|\chi(w - D_N \overline{w})| = O(\chi \delta^{2N+2}) = O(\delta^{2N+4/3}).$$

5.3   Time Relaxation Does Not Alter Shock Speeds

Time relaxation originated in the work of Schochet, Roseneau and Tadmor studying regularizations of shocks in conservation laws and was used for calculation of complex flows including shocks in the work of Stolz, Adams and Kleiser. In extensive tests, this time relaxation operator was seen to truncate the number of scales in the solutions of the resulting continuum model so as to be resolvable on a given mesh without affecting important solution properties such as shock position and velocity. The aim of this section (which follows [ELN06]) is to give a mathematical explanation of why such shock velocities (and hence positions) are not altered. To begin, consider the $1-d$ conservation law for $u(x,t)$, $-\infty < x < \infty$, $t \geq 0$:

$$\frac{\partial u}{\partial t} + \frac{\partial}{\partial x} q(u) = 0 , \quad -\infty < x < \infty, \quad t > 0 , \qquad (5.8)$$

$$u(x,0) = u_0(x) , \quad -\infty < x < \infty, \qquad (5.9)$$

where $q(\cdot)$ denotes the flux function. Let $\delta > 0$ be given and for $u \in L^\infty(-\infty, \infty)$, the differential filtered value $\bar{u}$ is defined as the unique solution of

$$-\delta^2 \frac{\partial^2}{\partial x^2} \bar{u}(x) + \bar{u}(x) = u(x) . \qquad (5.10)$$

In this simple setting, the variation of constants gives an explicit formula for $\bar{u}(x)$ to be

$$\bar{u}(x) = \frac{1}{2\delta} \left\{ \exp^{x/\delta} \int_x^\infty \exp^{-y/\delta} u(y)\, dy \; + \; \exp^{-x/\delta} \int_{-\infty}^x \exp^{y/\delta} u(y)\, dy \right\}$$

$$= \frac{1}{2\delta} \int_{-\infty}^\infty \exp^{-|x-y|/\delta} u(y)\, dy \; . \tag{5.11}$$

Adding time relaxation to the conservation law changes the model and its solution of course. Thus we consider

$$\frac{\partial w}{\partial t} + \frac{\partial}{\partial x} q(w) + \chi\,(w - \overline{w}) = 0\; , \qquad -\infty < x < \infty, \quad t > 0\; , \tag{5.12}$$

$$w(x,0) = u_0(x)\; , \qquad -\infty < x < \infty. \tag{5.13}$$

We can associate with the conservation law plus time relaxation a modified flux function as follows and apply the theory of conservation laws to read off from it the shock speed. Indeed, from the definition of the filter

$$-\delta^2 \frac{\partial}{\partial x^2} \overline{w} = \; w - \overline{w}\; ,$$

and thus

$$\chi\,(w - \overline{w}) = \; \frac{\partial}{\partial x} \left(-\chi \delta^2\, \overline{w}_x \right)\; .$$

The model (5.12) can then be rewritten as

$$\frac{\partial w}{\partial t} + \frac{\partial}{\partial x} q_\delta(w) = \; 0\; , \quad \text{where } q_\delta(w) := q(w) - \chi\delta^2 \overline{w}_x\; . \tag{5.14}$$

Consider the (standard) case of a discontinuous initial condition

$$u_0(x) = \begin{cases} u_L & x < 0, \\ u_R & x \geq 0. \end{cases} \tag{5.15}$$

The theory of conservation laws implies that when a shock develops in the original conservation law the true shock velocity, U, is given by

$$U = \frac{q(u_L) - q(u_R)}{u_L - u_R}\; . \tag{5.16}$$

The same theory applied to $\partial w/\partial t + \partial q_\delta(w)/\partial x = 0$ gives the shock velocity in the model's solution, W, as

$$W = \frac{q_\delta(u_L) - q_\delta(u_R)}{u_L - u_R}\; . \tag{5.17}$$

Comparing and using the formula for $q_\delta(\cdot)$ we conclude $U = W \Leftrightarrow [q] = [q_\delta]$ where $[\cdot]$ denotes the jump across the shock. We calculate

$$[q] - [q_\delta] = \chi \delta^2 (\bar{w}_x|_L - \bar{w}_x|_R) = \chi \delta^2 [\bar{w}_x].$$

The correct shock speed will be predicted correctly if, for piecewise continuous w, $[\bar{w}_x] = 0$, i.e., if $\bar{w}_x$ is continuous! Conversely, if $\bar{w}_x$ is discontinuous across a discontinuity in w, $[q_\delta] \neq [q]$ and the shock velocity will be altered and thus the shock position as well. For the differential filter $\bar{w}_x$ is continuous. (Indeed, this follows from either directly from the formula for $\bar{w}$ or by noting $\bar{w}_x$ is a primitive of the piecewise continuous function $\bar{w}_{xx}$ and is thus continuous.) For averaging by convolution, $\bar{w}$ will have the same smoothness as the filter kernel. Thus, $[\bar{w}_x] = 0$ provided the filter kernel $g(x)$ is at least $C^1(\mathbb{R})$.

We conclude that *shock speeds are unchanged when time relaxation is added to a conservation law if filtering is done with a second order differential filter or by convolution with a C^1 filter kernel. Averaging by convolution with a (C^0) top-hat filter might alter the model's shock speeds!*

5.4   Nonlinear Time Relaxation

The simplest example of such a models arises when no closure model is used on the nonlinear term and the time relaxation regularization is used

$$w_t + \nabla \cdot (w\ w) - \nu \triangle w + \nabla q + \chi w^* = f, \ and \ \nabla \cdot w = 0. \qquad (5.18)$$

The time relaxation term χw^* is a type of frictional force which damps fluctuations. However,

- *It does not have units of friction since wet friction is quadratic and scales like $\chi |w^*| w^*$.*
- *linear friction χw^* is local in a sense in scale space but global in physical space. Nonlinear friction can be local in both scale and physical space.*

This section therefore considers the natural regularization of nonlinear friction:

$$w_t + \nabla \cdot (w\ w) - \nu \triangle w + \nabla q + \chi |w^*| w^* = f, \ and \ \nabla \cdot w = 0. \qquad (5.19)$$

The simplest $(N = 0)$ case is

$$w_t + \nabla \cdot (w\ w) - \nu \triangle w + \nabla q + \chi |w - \bar{w}|(w - \bar{w}) = f, \ and \ \nabla \cdot w = 0. \ (5.20)$$

In the nonlinear case, the parameter has units $[\chi] = 1/length$. Even in the $N = 0$ case above there are many interesting open questions. *The main goal of this section is to delineate some of the interesting open questions in nonlinear time relaxation.*

5.4.1 Open Question 1: Does Nonlinear Time Relaxation Dissipate Energy in All Cases?

We know that linear friction dissipates energy since

$$\int_\Omega \chi(w - \overline{w}) \cdot w \, dx = \int_\Omega \chi(-\delta^2 \triangle \overline{w}) \cdot w \, dx =$$

$$= \chi \delta^2 \int_\Omega \nabla w \cdot (-\delta^2 \triangle + 1)^{-1} \nabla w \, dx \geq 0$$

since $(-\delta^2 \triangle + 1)^{-1}$ is a self adjoint and positive operator. (Alternately, it can be shown by expanding w in a Fourier series, inserting and calculating.)

The question of energy dissipation of nonlinear time relaxation is mathematically equivalent to asking: if:

$$\int_\Omega |w - \overline{w}|(w - \overline{w}) \cdot w \, dx \geq 0, \forall w \in L^2(\Omega)? \tag{5.21}$$

If this fails for simple filters, the question can be reformulated as asking if there must be some compatibility condition linking filtering and nonlinear time relaxation to ensure positivity of the above.

5.4.2 Open Question 2: If not, What Is the Simplest Modification to Nonlinear Time Relaxation that Always Dissipates Energy?

5.4.3 Open Question 3: How Is Nonlinear Time Relaxation to be Discretized in Time so as to be Unconditionally Stable and Require Filtering Only of Known Functions?

The most natural attempt is, suppressing spacial discretizations, to time step by

$$\frac{w^{n+1} - w^n}{\triangle t} + w^n \cdot \nabla w^{n+1} - \nu \triangle w^{n+1} + \nabla p^{n+1} + \tag{5.22}$$

$$+ \chi |w^n - \overline{w}^n|(w^{n+1} - \overline{w}^n) = f^{n+1} \tag{5.23}$$

$$\text{and } \nabla \cdot w^{n+1} = 0. \tag{5.24}$$

It is not entirely clear if this method is stable. It seems possible that the time discretization of the relaxation term might input nonphysical numerical energy into the approximate solution. A better attempt is to adapt ideas of Connors, Howell, and WL [CHL09] and try instead:

$$\frac{w^{n+1} - w^n}{\triangle t} + w^n \cdot \nabla w^{n+1} - \nu \triangle w^{n+1} + \nabla p^{n+1} \tag{5.25}$$

$$+ \chi |w^n - \overline{w}^n| w^{n+1} - \chi \sqrt{|w^n - \overline{w}^n| |w^{n-1} - \overline{w}^{n-1}|} \overline{w}^n = f^{n+1} \tag{5.26}$$

$$\text{and } \nabla \cdot w^{n+1} = 0. \tag{5.27}$$

The challenge is now to show that one of the above (or a third option) is stable.

5.5 Analysis of a Nonlinear Time Relaxation

Nonlinear time relaxation can be modified to ensure energy dissipation in all cases. This modification is developed next. It is important to note that the modification does not reduce the interest in the above three questions because the modification requires substantially more storage and floating point operations per time step than the unmodified nonlinear time relaxation term. Thus, in this section, we change the relaxation term to make the energy dissipation amenable to the meager mathematical tools available for nonlinear problems. Beginning with the simplest $N = 0$ case, the modified model we consider is

$$w_t + \nabla \cdot (w \ w) - \nu \triangle w + \nabla q + \chi (I - G) \left[|w - \overline{w}| (w - \overline{w}) \right] = f, \text{ and } \nabla \cdot w = 0. \tag{5.28}$$

The motivation for the change is that now we can guarantee that energy is dissipated by the nonlinear relaxation term since

$$\int_\Omega \chi (I - G) \left[|w - \overline{w}| (w - \overline{w}) \right] \cdot w dx$$

$$= \int_\Omega \chi \left[|w - \overline{w}| (w - \overline{w}) \right] \cdot (I - G) w dx$$

$$= \int_\Omega \chi |w - \overline{w}| (w - \overline{w}) \cdot (w - \overline{w}) dx$$

$$= \int_\Omega \chi |w - \overline{w}|^3 dx \geq 0.$$

The problem is that the extra $(I - G)$ makes the problem of stable; time stepping is more difficult and adds more filtering steps and thus cost. For example, solution of this modified equation by time stepping likely requires defining additional equations all coupled together by:

$$w_t + \nabla \cdot (w\ w) - \nu \triangle w + \nabla q + \chi \left[|w - \phi|(w - \phi) - \psi \right] = f, \quad (5.29)$$

$$(-\delta^2 \triangle + 1)\phi = w \qquad\qquad\qquad\qquad\qquad\qquad (5.30)$$

$$(-\delta^2 \triangle + 1)\psi = |w - \phi|(w - \phi) \qquad\qquad\qquad\qquad (5.31)$$

$$\text{and } \nabla \cdot w = 0. \qquad\qquad\qquad\qquad\qquad\qquad\qquad (5.32)$$

If this is discretized with N_V velocity and N_P pressure degrees of freedom (per time step), the coupled system, due to the extra variables introduced, has $3N_V + N_P$ degrees of freedom.

5.5.1   Open Question 4: Is the Extra (I − G) Necessary to Ensure Energy Dissipation or Just a Mathematical Convenience?

5.5.2   Open Question 5: If the Extra (I − G) Is Necessary, How Is it to be Discretized in Time?

In the general case, the modified nonlinear relaxation model we consider is to find a L-periodic (with zero mean) velocity and pressure satisfying

$$w(x,0) = u_0(x), \text{ in } \Omega,$$

$$\nabla \cdot w = 0, \text{ in } \Omega \times (0, T),$$

$$w_t + w \cdot \nabla w + \nabla q + \nu \triangle w$$

$$+\chi(I - D_N G)\left[|w - D_N \overline{w}|(w - D_N \overline{w}) \right] = f, \text{ in } \Omega \times (0, T) . \quad (5.33)$$

The relaxation coefficient χ must be specified .

Proposition 47. *Let $u_0 \in L_0^2(\Omega), f \in L^2(\Omega \times (0, T))$, and $\int_\Omega f(x, t)dx = 0$. For $\delta > 0$, let the averaging be $(-\delta^2 \triangle + 1)^{-1}$. If w is a strong solution of the model (5.33), w satisfies*

$$\frac{1}{2}\|w(t)\|^2 + \int_0^t \int_\Omega \nu |\nabla w|^2 + \chi |w - D_N \overline{w}|^3 dxdt' = \frac{1}{2}\|u_0\|^2 + \int_0^t \int_\Omega f \cdot w dxdt'.$$

Proof. Multiply the equation by w, and integrate over the domain Ω. Use

$$\int_\Omega \chi(I - DG)\left[\|w - D\overline{w}\|(w - D\overline{w})\right] \cdot w\,dx$$

$$= \int_\Omega \chi\left[\|w - D\overline{w}\|(w - D\overline{w})\right] \cdot (I - DG)w\,dx$$

$$= \int_\Omega \chi|w - D\overline{w}|(w - D\overline{w}) \cdot (w - D\overline{w})\,dx$$

$$= \int_\Omega \chi|w - D\overline{w}|^3 dx \geq 0.$$

Finally integrate from 0 to t.   □

By the above energy estimate, the model's nonlinear relaxation term extracts energy from resolved scales. Thus, define an energy dissipation rate induced by time relaxation as

$$\varepsilon_{model}(w)(t) := \frac{1}{|\Omega|}\int_\Omega \nu|\nabla w|^2 + \chi|w - D_N\overline{w}|^3 dx. \tag{5.34}$$

Provided viscous dissipation is negligible we can approximate this by

$$\varepsilon_{model}(w)(t) \simeq \frac{1}{|\Omega|}\int_\Omega \chi|w - D_N\overline{w}|^3 dx \tag{5.35}$$

When $N = 0$ we have

$$\varepsilon_{model}(w)(t) \simeq \frac{1}{|\Omega|}\int_\Omega \chi|w - \overline{w}|^3 dx = \tag{5.36}$$

$$= \frac{1}{|\Omega|}\int_\Omega \chi|\delta^2\triangle\overline{w}|^3 dx = \frac{\chi\delta^6}{|\Omega|}\int_\Omega |\triangle(-\delta^2\triangle + 1)^{-1}w|^3 dx. \tag{5.37}$$

The model's kinetic energy is the same as for the Navier–Stokes equations

$$E_{model}(w)(t) := \frac{1}{2}\|w(t)\|^2. \tag{5.38}$$

Proceeding under K41 phenomenology, the ratio of nonlinearity to dissipative effects is

$$R_N \simeq \frac{|w \cdot \nabla w|}{\chi|(I - D_N G)|w - D_N Gw|(w - D_N Gw)|}.$$

5.5.3 The N=0 Case

For example, if $N = 0$, and keeping in mind that for the large scales $(\frac{\delta}{L})^2 << 1$, then we have

$$R_0 \simeq \frac{|w \cdot \nabla w|}{\chi|(I - G)||(w - \overline{w})|^2} = \frac{|w \cdot \nabla w|}{\chi|\delta^2 \triangle||\delta^2 \triangle \overline{w}|^2} =$$

$$\simeq \frac{|w \cdot \nabla w|}{\chi|\delta^2 \triangle||\delta^2 \triangle(-\delta^2 \triangle + 1)^{-1}w|^2} \simeq \frac{U \frac{1}{L} U}{\chi \left(\frac{\delta}{L}\right)^2 |\left(\frac{\delta}{L}\right)^2 (\frac{\delta^2}{L^2} + 1)^{-1}U|^2} =$$

$$= \frac{\frac{1}{L}}{\chi \left(\frac{\delta}{L}\right)^6 |(\frac{\delta^2}{L^2} + 1)^{-1}|^2} = \frac{1}{L\chi} \left(\frac{\delta}{L}\right)^{-6} (\frac{\delta^2}{L^2} + 1)^2$$

$$\simeq \frac{L^5}{\chi \delta^6}$$

Definition 48. The non-dimensionalized time relaxation parameter for the Navier–Stokes equations + nonlinear time relaxation is

$$R_0 = \frac{L^5}{\chi \delta^6}, \text{ for } N = 0.$$

Next we must form the small scales parameter which measures the ratio of nonlinearity to dissipation at the smallest persistent scales. Let w_{small} denote a characteristic velocity of the smallest persistent eddies and let η_{model} denote the length scale associated with them. For the small scales it is no longer reasonable to suppose δ is small with respect to η_{model}. Then, we calculate

$$R_{0-small} \simeq \frac{|w \cdot \nabla w|}{\chi|(I - G)||(w - \overline{w})|^2} = \frac{|w \cdot \nabla w|}{\chi|\delta^2 \triangle||\delta^2 \triangle \overline{w}|^2} =$$

$$= \frac{|w \cdot \nabla w|}{\chi|\delta^2 \triangle||\delta^2 \triangle(-\delta^2 \triangle + 1)^{-1}w|^2}$$

$$\simeq \frac{w_{small} \frac{1}{\eta} w_{small}}{\chi \left(\frac{\delta}{\eta}\right)^2 |\left(\frac{\delta}{\eta}\right)^2 (\frac{\delta^2}{\eta^2} + 1)^{-1}w_{small}|^2}$$

$$= \frac{\frac{1}{\eta}}{\chi \left(\frac{\delta}{\eta}\right)^6 |(\frac{\delta^2}{\eta^2} + 1)^{-1}|^2} = \frac{1}{\eta \chi} \left(\frac{\delta}{\eta}\right)^{-6} \left(\frac{\delta^2}{\eta^2} + 1\right)^2.$$

The estimate of the smallest resolved scales is based upon two principles that $R_{small} = O(1)$ at length-scale η_{model} and statistical equilibrium. The power input at the large scales like $\frac{U^3}{L}$. The dissipation at the smallest resolved scales is

$$\text{small scale dissipation rate} \simeq \chi |w - \overline{w}|^3 \simeq$$

$$\simeq \chi |\delta^2 (\triangle \overline{w})|^3$$

$$\simeq \chi \left(\frac{\delta^2}{\eta_{model}^2} \right)^3 \left[\left(1 + \frac{\delta^2}{\eta_{model}^2} \right) w_{small} \right]^3 .$$

These two conditions thus give the pair of equations

$$\frac{\eta^5}{\chi \delta^6} \left(\frac{\delta^2}{\eta^2} + 1 \right)^2 \simeq 1 \text{ , and}$$

$$\frac{U^3}{L} \simeq \chi \frac{\delta^6}{\eta_{model}^6} \left(1 + \frac{\delta^2}{\eta_{model}^2} \right)^3 w_{small}^3 .$$

These are the fundamental equation determining the model's micro-scale.

Case: Perfect resolution In this case $\delta = \eta_{model}$ so that $1 + \frac{\delta^2}{\eta_{model}^2} \simeq 2$.

In this case the interesting question is to determine the choice of relaxation parameter that enforces this case. The above two determining equations become

$$\frac{4}{\chi \delta} \simeq 1 \text{ , and}$$

$$\frac{U^3}{L} \simeq 8 \chi w_{small}^3 .$$

Solving gives

$$\chi \simeq 4\delta^{-1} \text{ and } w_{small} = \left(\frac{1}{2\sqrt[3]{2}} \sqrt[3]{\frac{\delta}{L}} \right) U.$$

Open Question 7: Come to an understanding of the reason we can determine the microscale in this case only through $\mathbf{R} = \mathbf{1}$, *without using energy in = energy out*

5.5.4   The Analysis of Lilly

For the modified nonlinear time relaxation term, the derivation of Lilly for the Smagorinsky model, well known in LES, e.g., Sagaut [S01], can be repeated for the above. This derivation is heuristic but gives a useful indication of the scaling of the relaxation parameter with respect to the other model parameters. The idea of Lilly is to equate $\varepsilon = \varepsilon_{model}$ and evaluate the ε_{model} by assuming (among other things) that the velocity field arises from homogeneous isotropic turbulence, $E(k) \simeq \alpha \varepsilon^{\frac{2}{3}} k^{-\frac{5}{3}}$. To use energy spectrum information a further assumption is needed that the following two are of comparable orders of magnitude:

$$< \chi ||w - \overline{w}||^3_{L^3(\Omega)} > \simeq \chi < ||w - \overline{w}||^2_{L^2(\Omega)} >^{\frac{3}{2}} .$$

With this assumption we can calculate

$$\varepsilon_{model} = \chi \left[\int_{k_{min}}^{k_{max}} \left(1 - \frac{1}{\delta^2 k^2 + 1} \right)^2 E(k) dk \right]^{\frac{3}{2}} , \text{ where}$$

$$E(k) \simeq \alpha \varepsilon^{\frac{2}{3}} k^{-\frac{5}{3}} , \ \alpha = \text{Kolmogorov constant.}$$

Thus we have

$$\varepsilon_{model} = \chi \left[\int_{k_{min}}^{k_{max}} \left(1 - \frac{1}{\delta^2 k^2 + 1} \right)^2 \alpha \varepsilon^{\frac{2}{3}} k^{-\frac{5}{3}} dk \right]^{\frac{3}{2}}$$

$$= \chi \alpha^{\frac{3}{2}} \varepsilon \left[\int_{k_{min}}^{k_{max}} \left(\frac{\delta^2 k^2}{\delta^2 k^2 + 1} \right)^2 k^{-\frac{5}{3}} dk \right]^{\frac{3}{2}}$$

$$= \chi \alpha^{\frac{3}{2}} \varepsilon \delta^{\frac{5}{2} - \frac{3}{2}} \left[\int_{k_{min}}^{k_{max}} \left(\frac{\delta^2 k^2}{\delta^2 k^2 + 1} \right)^2 (\delta k)^{-\frac{5}{3}} \delta dk \right]^{\frac{3}{2}}$$

With the change of variable $z = \delta k$ this reduces to

$$\varepsilon_{model} = \chi \alpha^{\frac{3}{2}} \varepsilon \delta \left[\int_{z_{min}}^{z_{max}} \left(\frac{z^2}{1 + z^2} \right)^2 z^{-\frac{5}{3}} dz \right]^{\frac{3}{2}} .$$

Setting $\varepsilon = \varepsilon_{model}$ thus gives the following value of the relaxation parameter (after simplification)

$$\chi = [\alpha^{\frac{3}{2}} \delta \beta_0]^{-1} , \text{ where}$$

$$\beta_0 = \int_{z_{min}}^{z_{max}} \left(\frac{z^2}{1 + z^2} \right)^2 z^{-\frac{5}{3}} dz.$$

5.5.5 Open Question 8: Is It Possible to Extend the Above Calculation of the Optimal Relaxation Parameter to the Original Version of Nonlinear Time Relaxation?

5.6 Remarks

This chapter is based on [LN06a]. For further developments on the numerical analysis of time relaxation methods see [L07b, Gue04, CL10, ELN06], and [N10]. There is a related and very promising approach (with many open mathematical questions) of TLES: LES based on time filtering and time relaxation. See [LPR10, ABHM06, DB86, Max79, Pr06, PTGG06] and [PGGT03] for this work. Time relaxation, using deconvolution operators, was also introduced into LES by Stolz, Adams, and Kleiser in their early papers. For early work on time relaxation as a continuum model see [R89, ST92].

The theory shows that the effective cutoff actually decreases as both $\delta \to 0$ for fixed N and (more slowly) as $N \to \infty$ for fixed δ. This suggests that as N increases one should simultaneously slowly increase δ so as to keep a fixed effective cutoff length. This effect is possibly reflected in the dependency of $\chi_{optimal}$ upon N. To our knowledge, it has not been tested.

Chapter 6
The Leray-Deconvolution Regularization

6.1 The Leray Regularization

"In all things, be prepared to critique the model's assumptions and compare predictions to actual events."

G. Washington.

In *LES* one solves a system whose solution is an approximation to local spacial averages of the NSE. In *regularization modeling* one solves a system similar to the NSE which has better qualitative properties for numerical simulation than the underlying NSE. As a consequence, the computed solution is simply a regularized approximation of the NSE solution rather than a local, spacial average of the fluid velocity. The many possible NSE regularizations can be judged based on:

(a) Accuracy as an approximation of the NSE.
(b) Fidelity to qualitative properties of the NSE's velocity.
(c) Ease of solution with standard numerical methods and packages for flow problems.

In 1934 J. Leray [L34a, L34b] studied an interesting regularization of the Navier–Stokes equations. He proved that the *regularized* Navier–Stokes equations has a unique, smooth, strong solution and that as the regularization length-scale $\delta \to 0$, the regularized system's solution converges (modulo a subsequence) to a weak solution of the Navier–Stokes equations. This model has recently been attracting new interest as a continuum model upon which large eddy simulation can be based (see, e.g., Geurts and Holm [GH03] and [CHOT05, VTC05]). If $\overline{w}$ denotes a local spacial average of the velocity w associated with filter length-scale δ, the classical Leray model is given by

W.J. Layton and L.G. Rebholz, *Approximate Deconvolution Models of Turbulence*, Lecture Notes in Mathematics 2042, DOI 10.1007/978-3-642-24409-4_6, © Springer-Verlag Berlin Heidelberg 2012

$$\frac{\partial w}{\partial t} + \overline{w} \cdot \nabla w - \nu \triangle w + \nabla q = f, \qquad (6.1)$$

$$\nabla \cdot w = 0. \qquad (6.2)$$

Leray chose $\overline{w} = g_\delta \star w$, where g_δ is a Gaussian[1] with averaging radius δ:

$$\overline{w}(x,t) := g_\delta \star w(x,t), \text{ where}$$

$$g_\delta(x) := \delta^{-3} g(x/\delta), \text{ and } g(x) := Gaussian,$$

$$g_\delta \star w(x,t) := \int_{\mathbb{R}^3} g_\delta(x') w(x - x', t) dx'.$$

The simplest context to study any flow model is periodic BCs: we take $\Omega = (0, L)^3$, the initial condition

$$w(x, 0) = \overline{w}_0(x), \text{ in } \Omega ,$$

and impose periodic boundary conditions (with zero mean) on the solution (and all problem data)

$$w(x + Le_j, t) = w(x, t), \text{ and } \int_\Omega w(x, t) dx = 0.$$

One way to try to resolve the existence and uniqueness question is indirectly. The equations are smoothed just enough that solutions exist, are regular, and unique. Then as the regularization parameter approaches zero, J. Leray also considered the solution of the fundamental problem. In the next theorem, Leray was unable to eliminate the possibility that different subsequences might converge to different weak solutions however.

Theorem 49 (Leray 1934). *Let the initial condition, body force and the solution's boundary conditions be periodic with zero mean. Let*

$$\overline{u} := (-\delta^2 \Delta + 1)^{-1} u .$$

A unique strong solution to the Leray model exists and converges (modulo a subsequence) to a weak solution of the NSE as the averaging radius $\delta_j \to 0$.

If $\nu = f = 0$ (and under periodic boundary conditions): Leray's model exactly conserves kinetic energy. If the NSE has a unique and smooth enough strong solution then additionally

$$\sup_{0 \le t \le T_{final}} ||u_{NSE} - w_{Leray}|| \le C(problem\_data)\delta^2.$$

---

[1] By other choices of convolution kernel, differential filters, sharp spectral cutoff and top-hat filters can be recovered.

Proof. See [L34a, L34b] for the case of a Gaussian filter. The case of a differential filter follows by the same arguments. Conservation of energy follows by taking the inner product with w_{Leray} and integrating. The error estimate follows by subtraction, integration and applying Gronwall's' inequality.   □

These and other good theoretical properties have sparked a re-examination of the Leray model for simulations of turbulent flows [Reb07] [LL08] [GH05] [D03] [CHOT05]. The most common modification is that the Gaussian filter is replaced by a less expensive differential filter,

$$\overline{u} := (-\delta^2 \Delta + 1)^{-1} u.$$

The form of the model, its theory and the tests of Geurts and Holm [GH05] reveal three issues:

1. $\overline{u}$ is a nonlocal function of u and so must not be treated implicitly in any numerical simulation.
2. The accuracy[2] of the Leray model using differential filters is strictly limited to $O(\delta^2)$ at best.
3. Without additional terms added, simulations of the model can result in an accumulation of energy around the cutoff length scale (i.e. wiggles) because the model's cutoff is very much smaller than the filter radius.

These deficiencies have sparked extension of the Leray regularization to higher accuracy ones (the Leray-deconvolution family of models) and ones with more physical fidelity (the NS-α model). These are very promising but will likely not be the last word in the development of this early idea of Jean Leray. In studying different regularizations among the infinitude of possible ones, it is important to ask :

What are the requirements for a regularization to be a good one?

- Accuracy in smooth regions is certainly very important.
- Physical fidelity in the sense of conserving as many important integral invariants as possible is also desired.
- The third critical condition is control over the regularization's cutoff length scale: it should be $O(filter\ radius)$ for a successful regularization:

$$\eta_{model} = O(\delta) \Rightarrow Success!$$
$$\eta_{model} < O(\delta) \Rightarrow Failure!$$

---

[2]Here "accuracy" is meant in the most favorable sense: The consistency error of the model evaluated for $C^\infty(\Omega \times (0,T))$ solutions of the NSE is $O(\delta^2)$ and no better.

6.2 Dunca's Leray-Deconvolution Regularization

The Leray model is easy to solve using standard numerical methods for the Navier–Stokes equations, e.g., [LMNR08b], and, properly interpreted, requires no extra or ad hoc boundary conditions in the non-periodic case. However, as a model upon which to base a numerical simulation it has shortcomings:

- The Leray model's solution can suffer an accumulation of energy around the cutoff frequency or filter length scale or meshwidth, [GH03].
- $\overline{w}$ is a nonlocal function of u and so must not be treated implicitly.
- The model has low accuracy on the smooth flow components, e.g., $\|w_{NSE} - w\| = O(\delta^2)$.
- It does not contain enough dissipation to prevent non-physical accumulation of energy around the cutoff length-scale. In other words, its microscale is significantly smaller than the filter radius. Simulations, without additional terms added, result in an accumulation of energy around the cutoff length scale (i.e. wiggles).
- It is not frame invariant, Guermond, Prudhomme and Oden [GOP03].
- It does not conserve helicity in 3d.

Further, there is the vexing question of

- What does the model's solution mean, and in what sense can it be considered accurate?

The drawback of accumulation of energy at the cut-off length scale can be overcome by various numerical devices including time relaxation [Gue04, LN06a, SAK01a, SAK01b, SAK02].

Finally, there is the issue of the accuracy upon the large scales/smooth flow components which must be improved if it is to evolve from a descriptive regularization to a predictive model. The consistency error of the model, considered as an approximation of the Navier–Stokes equations, is

$$\tau_{Leray} = \overline{u}u - uu.$$

Since $\tau_{Leray} = (\overline{u} - u)u = O(\delta^2)$, for smooth velocity fields, the model is of quite low accuracy. This low accuracy is overcome by a clever idea of A. Dunca which is to reformulate the model into a deconvolution model:

$$w_t + D(\overline{w}) \cdot \nabla w - \nu \triangle w + \nabla q = f, \quad \nabla \cdot w = 0.$$

The Leray-deconvolution models have arbitrarily high orders of accuracy, include the Leray-alpha model as the zeroth order ($N = 0$) case, and have the following attractive properties:

- For $N = 0$ they include the Leray/Leray-α model as the lowest order special case.
- Their accuracy is high, $O(\delta^{2N+2})$ for arbitrary $N = 0, 1, 2, \ldots$.
- They improve upon the attractive theoretical properties of the Leray model, e.g. convergence (modulo a subsequence) as $\delta \to 0$ to a weak solution of the NSE and $\|u_{NSE} - u_{LerayDCM}\| = O(\delta^{2N+2})$ for a smooth, strong solution u_{NSE}, [LL08].
- Given $\overline{u}$ the computation of $D_N \overline{u}$ is computationally attractive.
- The higher order models (for $N \geq 1$) give dramatic improvement of accuracy and physical fidelity over the $N = 0$ case.
- Increasing model accuracy can be done in two ways: (a) cutting $\delta \to \delta/2$ increases accuracy for $N = 0$ by approximately a factor of 4 but requires remeshing with approximately 8 times as many unknowns, and (b) increasing $N \to N+1$ increases accuracy from $O(\delta^{2N+2})$ to $O(\delta^{2N+4})$ and requires one more Poisson solve $((-\delta^2 \Delta + 1)^{-1}\phi)$ per time step.

6.3 Analysis of the Leray-Deconvolution Regularization

6.3.1 Existence of Solutions

The theory of Leray-deconvolution model begins, like the Leray theory of the Navier–Stokes equations, with a clear global energy balance, and existence and uniqueness of solutions. Let $w_0 \in L_0^2 \cap \{v : \nabla \cdot v = 0\}$, and $f \in H^{-1}$. The problem we consider is the following, for a fixed $T > 0$, find (w, q) satisfying

$$w \in L^2([0, T], H^1_{periodic} \cap \{v : \nabla \cdot v = 0\}), \tag{6.3}$$

$$w \in L^\infty([0, T], L_0^2 \cap \{v : \nabla \cdot v = 0\}), \tag{6.4}$$

$$\frac{\partial w}{\partial t} \in L^2([0, T], H^{-1}) \tag{6.5}$$

$$q \in L^2([0, T], L_0^2), \tag{6.6}$$

and in the distributional sense,

$$\frac{\partial w}{\partial t} + D_N(\overline{w}) \cdot \nabla w - \nu \triangle w + \nabla q = D_N \overline{f}, \tag{6.7}$$

$$w(x, 0) = D_N \overline{w_0} = w_0, \tag{6.8}$$

where L_0^2 denotes the functions, vector or scalar, in $L^2(0, 2\pi)$ with zero mean value.

Theorem 50. *The Leray deconvolution model admits a unique solution* (w, q), *where* w *satisfies the energy equality*

$$\frac{1}{2}\|w(t)\|^2 + \nu \int_0^t \int_\Omega |\nabla w|^2 dx dt' = \frac{1}{2}\|D_N \overline{w_0}\|^2 + \int_0^t \int_\Omega D_N \overline{f} \cdot w \, dx dt'. \quad (6.9)$$

Moreover,

$$w \in L^\infty([0,T], H^1_{periodic} \cap \{v : \nabla \cdot v = 0\}) \cap L^2([0,T], H^2).$$

6.3.2   Proof

For simplification, denote by $\|\cdot\|_{p,B}$ the norm in $L^p([0,T], B)$ and when $B = H^s$, $\|\cdot\|_{p,s}$ for simplicity. Consider also the space

$$V = L^2([0,T], H^1_{periodic} \cap \{v : \nabla \cdot v = 0\}) \cap L^\infty([0,T], L^2_0 \cap \{v : \nabla \cdot v = 0\}),$$

$$\|w\|_{\mathbf{V}} = \|w\|_{2,1} + \|w\|_{\infty,0}.$$

Recall that

$$\forall w \in V, \quad \forall r \in [2,6], \quad \|w\|_{\frac{4r}{3(r-2)}, L^r} \leq C\|w\|_V. \quad (6.10)$$

We introduce the operators:

$$A(w,v) = \nu \int_0^T \int_\Omega \nabla w : \nabla v dx dt,$$

$$B_N(w,v) = \int_0^T \int_\Omega (D_N(\overline{w}) \cdot \nabla) w \cdot v dx dt = -\int_0^T \int_\Omega D_N(\overline{w}) \otimes w : \nabla v dx dt.$$

We first notice that

$$|A(w,v)| \leq \nu \|w\|_{2,1} \|v\|_{2,1}. \quad (6.11)$$

Now assume that $w \in V$, $v \in L^2([0,T], H^1_{periodic} \cap \{v : \nabla \cdot v = 0\})$. It is known that there exists a constant C which depends on N and such that

$$|B_N(w,v)| \leq C\|w\|_V^2 \|v\|_{2,1}, \quad B_N(w,w) = 0. \quad (6.12)$$

Here we have remarked that because $H^2 \subset (L^\infty(\mathbb{R}))^3$, $w \in L^\infty([0,T], L^2_0 \cap \{v : \nabla \cdot v = 0\})$ yields $D_N G(w) \in L^\infty([0,T], H^2)$ and therefore,

$$D_N(\overline{w}) \in L^\infty([0,T] \times \mathbb{R}^3), \quad \|w\|_{\infty, L^\infty} \leq C\|w\|_V \quad (6.13)$$

for a constant C which depends on N. Define

$$W = \{w \in V, \; \frac{\partial w}{\partial t} \in L^2([0,T], H^{-1})\}. \tag{6.14}$$

Notice that any $w \in W$ is almost everywhere equal to a function in $C^0([0,T], L_0^2 \cap \{v : \nabla \cdot v = 0\})$. Therefore,

$$W \subset C^0([0,T], L_0^2 \cap \{v : \nabla \cdot v = 0\}).$$

Finally, when $f \in L^2([0,T], H^{-1})$, $D_N \overline{f} \in L^2([0,T], H^1_{periodic} \cap \{v : \nabla \cdot v = 0\})$ the variational formulation is: Find $w \in W$ such that $w(0) = w_0$ and $\forall v \in L^2([0,T], H^1_{periodic} \cap \{v : \nabla \cdot v = 0\})$,

$$\int_0^T < \frac{\partial w}{\partial t}, v > dt + B_N(w,v) + A(w,v) = \int_0^T \int_\Omega D_N \overline{f} \cdot v \, dx dt.$$

The existence of a solution can be proven by the classical Galerkin approach, [LLe06]. By taking $w = v$, one directly gets the energy estimate. Notice that thanks to the energy equality and $||D_N(\overline{w})|| \le ||w||$, $||D_N \overline{f}||_{-1} \le ||f||_{-1}$, w satisfies the following estimates:

$$||w||^2_{\infty,0} \le ||w_0||^2 + \frac{1}{\nu}||f||^2_{2,-1}, \tag{6.15}$$

$$||w||^2_{2,1} \le \frac{1}{\nu}||w_0||^2 + \frac{1}{\nu^2}||f||^2_{2,-1}$$

in other words,

$$||w||_V \le C(\nu, ||w_0||, ||f||_{2,-1}). \tag{6.16}$$

The bound here does not depend on N. Notice also that we can derive an estimate for $\frac{\partial w}{\partial t}$ in the space $L^2([0,T], H^{-1})$. However the bound for $\frac{\partial w}{\partial t}$ in this space depends on N.

The pressure q is recovered thanks the De Rham Theorem and its regularity results from the fact that $\nabla q \in L^2([0,T], H^{-1})$.

We now check regularity. Let ∂w and ∂q be a derivative of w and q. Thanks to the periodic boundary conditions, one has

$$\partial_t \partial w + (D_N \overline{w} \nabla) \partial w - \nu \Delta \partial w + \nabla \partial q = \partial D_N \overline{f} - (\partial D_N \overline{w} \nabla) \cdot w. \tag{6.17}$$

One notes that

$$\partial D_N \overline{f} \in L^2([0,T], (L_0^2 \cap \{v : \nabla \cdot v = 0\})^3), \partial D_N \overline{w} \in L^\infty(0,T; H^1(\Omega)).$$

Since $w \in L^2([0,T], L_0^2 \cap \{v : \nabla \cdot v = 0\})$, by the Sobolev Imbedding Theorem, $(\partial D_N \overline{w} \nabla).w$ is periodic and in the space $L^2([0,T], (L^{3/2})^9)$. Since $1/6+2/3 = 5/6 < 1$, one has

$$(\partial D_N \overline{w} \nabla).w \in L^2([0,T], (L^{3/2})^9) \subset L^2([0,T], H^{-1}).$$

Then the equation admits a unique solution in the space V. By using the same technique as in [LLe06], it is easy to show that this solution is equal to ∂w showing that

$$w \in L^\infty([0,T], H^1_{periodic} \cap \{v : \nabla \cdot v = 0\}) \cap L^\infty([0,T], H^2).$$

In particular, there is some constant C (which depends on N) such that

$$\|w\|_{\infty,1} + \|w\|_{2,2} \leq C. \tag{6.18}$$

It remains to prove uniqueness. Let (w_1, q_1) and (w_2, q_2) be two solutions, $\delta w = w_1 - w_2$, $\delta q = q_2 - q_1$. Then one has

$$\partial_t \delta w + (D_N \overline{w_1} \nabla) \delta w - \nu \Delta \delta w + \nabla \delta q = (D_N \overline{\delta w} \nabla) w_2, \tag{6.19}$$

and $\delta w = 0$ at initial time. All the terms in the equation above being in $L^2([0,T], H^{-1})$, one can take $\delta w \in W \subset L^2([0,T], H^1_{periodic} \cap \{v : \nabla \cdot v = 0\})$ as test. Since $D_N \overline{w}$ is divergence free, one has

$$\int_0^T \int_\Omega (D_N \overline{w_1} \nabla) \delta w . \delta w = 0.$$

Meanwhile, one gets

$$\left| \int_\Omega (D_N \overline{\delta w} \nabla) w_2 . \delta w \right| \leq C \|D_N \overline{\delta w}\| \|\delta w\| \leq C \|\delta w\|^2$$

Therefore,

$$\frac{d}{2dt} \int \|\delta w\|^2 + \nu \int_\Omega |\nabla \delta w|^2 \leq C \|\delta w\|^2,$$

and we conclude that $\delta w = 0$ thanks to Gronwall's Lemma.

Remark 51. The Galerkin approximations to the solution w are under the form

$$w_n = \sum_{|k| \leq n} \widehat{w}_n(t, \mathbf{k}) e^{i\mathbf{k}.x}, \quad w_n(0, x) = \Pi_n(w_0)(x).$$

where the vector $(w_n(t, -n), ..., w_n(t, n))$ is a solution of an ODE and is of class C^1 on $[0, T]$ according to the Cauchy Lipschitz Theorem. We know on one hand that the sequence $(w_n)_{n\in\mathbb{Z}^+}$ converges strongly to w in $L^2([0,T),$ $L^2_0 \cap \{v : \nabla \cdot v = 0\})$ whether the sequence $(\frac{\partial w}{\partial t}\,_n)_{n\in\mathbb{Z}^+}$ converges weakly to $\frac{\partial w}{\partial t}$ in $L^2([0,T), H^{-1})$. On the other hand, for any smooth periodic field ϕ with $\nabla \cdot \phi = 0$ and $\phi(T,x) = 0$, since $w_n \in C^1([0,T], \mathbb{P}_n)$, integration by parts can be applied and gives

$$\int_0^T < \frac{\partial w}{\partial t}\,_n, \phi > dt = \int_0^T \int_\Omega \frac{\partial w}{\partial t}\,_n (t,x) \cdot \phi(t,x)\, dx dt =$$

$$\int_\Omega \Pi_n(w_0)(x) \cdot \phi(0,x)\, dx - \int_0^T \int_\Omega w_n(t,x) \cdot \partial_t \phi(t,x)\, dx dt$$

Passing to the limit when $n \to \infty$ yields

$$\int_0^T < \frac{\partial w}{\partial t}, \phi > dt = \int_\Omega w_0(x) \cdot \phi(0,x)\, dx - \int_0^T \int_\Omega w(t,x) \cdot \partial_t \phi(t,x)\, dx dt.$$
$$(6.20)$$

6.3.3 Limits of the Leray-Deconvolution Regularization

The models considered are intended as approximations to the Navier–Stokes equations. Thus the limiting behavior of the models solution is of primary interest. There are two natural limits: $\delta \to 0$ and $N \to \infty$. The first is the normal analytic question of turbulence modeling and considered already by J. Leray in 1934. The question of the behavior of the model's solution as $N \to \infty$ is much more unclear, however. In practical computations, cutting δ means re-meshing and increasing the memory and run time requirements greatly while increasing N simply means solving one more shifted Poisson problem per deconvolution step. Thus, increasing the accuracy of the model is much easier than decreasing its resolution. On the other hand, the van Cittert deconvolution procedure itself is an asymptotic approximation rather than a convergent one and has a large error at smaller length scales.

We denote by (w_N, q_N) the unique solution to the Leray-deconvolution problem, $w_N(0,x) = w_{0,N}(x) = D_N G(w_0)(x)$. The following was proven by Layton and Lewandowski [LL08].

Theorem 52. *Assume periodic with zero mean boundary conditions. There exists a subsequence N_j be such that w_{N_j} converges to a weak solution w to the Navier–Stokes equations. The convergence is weak in $L^2([0,T], H^1_{periodic} \cap \{v : \nabla \cdot v = 0\})$ and strong in $L^2([0,T], L^2_0 \cap \{v : \nabla \cdot v = 0\})$. For a fixed N, a subsequence of $w = w(\delta)$ converges to a weak solution of the Navier–Stokes solution when δ goes to zero.*

6.4   Accuracy of the Leray-Deconvolution Family

The accuracy of a regularization model as $\delta \to 0$ is typically studied in two ways. The first, called a posteriori analysis in turbulence model validation, is to obtain via direct numerical simulation (or from a DNS database) a "truth" solution of the Navier–Stokes equations, then, to solve the model numerically for varying values of δ and compute directly various modeling errors, such as $\overline{w} - w$. The second approach, known as á priori analysis in turbulence model validation studies (and is exactly an experimental estimation of a model's consistency error), is to compute the residual of the true solution of the Navier–Stokes equations (obtained from a DNS database) in the model. For example, to assess the consistency error of the Leray (and Leray-alpha model) model, the Navier–Stokes equations are rewritten to make the Leray model appear on the LHS as

$$\frac{\partial w}{\partial t} + \overline{w} \cdot \nabla w - \nu \triangle w + \nabla p - f = \nabla \cdot [\overline{w}w - ww] \quad \text{in } [0, T] \times \Omega.$$

The Leray-model's consistency error tensor is then $\tau_{Leray}(w, w) := \overline{w}w - ww$. Analysis of the modeling error in various deconvolution models, various norms and diverse settings in [LL08, BIL06] has shown that the energy norm of the model error, $\|\overline{u}_{NSE} - w_{model}\|$ or $\|u_{NSE} - w_{model}\|$ as appropriate, is driven by the consistency error tensor τ rather than $\nabla \cdot \tau$.

Thus, an analysis of a model's consistency error analysis evaluates $\|\overline{w}w - ww\|$. In the analysis of consistency errors, there are three interesting and important cases. Naturally, the case where w is a general, weak solution of the Navier–Stokes equations is most interesting and equally naturally, nothing can be expected within current mathematical techniques beyond very weak convergence to zero, possibly modulo a subsequence. Next is the case of smooth solutions (the classical case for evaluating consistency errors analytically). The case of smooth solutions is important for transitional flows and regions in non-homogeneous turbulence and it is an important analytical check that the LES model is very close to the Navier–Stokes equations on the large scales. The third case (introduced in [LL08]) is to study time averaged consistency errors using the intermediate regularity observed in typical time averaged turbulent velocities.

For the Gaussian filter it is known (e.g., Chap. 1 in [BIL06]) that for smooth ϕ, $\phi - (g_\delta \star \phi) = O(\delta^2)$ so that the Leray model's consistency error is second order accurate in δ on the smooth velocity components: $\tau_{Leray} = (\overline{w} - w)w = O(\delta^2)$. This simple calculation shows that the consistency error is dominated by the error in the regularization of the convecting velocity. Thus, improving the accuracy of a Leray-type regularization model hinges on improving the accuracy of the regularization. For the differential filter, from (1.2), $\phi - \overline{\phi} = \delta^2(-\triangle \overline{\phi})$ so the consistency error of the Leray-alpha model (the $N = 0$ and no time relaxation case of the Leray deconvolution family) is also $O(\delta^2)$.

We now consider the consistency error of the Leray-deconvolution model and show that the (asymptotic as $\delta \to 0$) consistency error is $O(\delta^{2N+2})$. To identify the consistency error tensor, the Navier–Stokes equations is rearranged

$$u_t + D_N(\overline{u}) \cdot \nabla u - \nu \triangle u + \nabla p - f = \nabla \cdot [D_N(\overline{u})u - uu], \text{ in } \Omega \times (0,T) \ .$$

The LHS is the Leray-deconvolution model and the RHS is the *residual of the true solution of the Navier–Stokes equations in the model*. Thus, the consistency error tensor is

$$\tau_N(u,u) := D_N(\overline{u})u - uu, \text{ for } N = 0, 1, \cdots .$$

The model error is driven by the model's consistency error $\tau_N(u,u)$ rather than $\nabla \cdot \tau_N(u,u)$. Since $\tau_N = (D_N(\overline{u}) - u)u$ the consistency error of the Leray-deconvolution model is dominated by the deconvolution error. As before, there are three cases: a general weak solution, solutions with the regularity typically observed in homogeneous, isotropic turbulence (see also [LL08]) and, to assess accuracy on the large scales, very smooth solutions. In this case the deconvolution error is bounded in [DE06, LL08] and induces a high order consistency error bound, given next.

Proposition 53. *The consistency error of the Nth Leray-deconvolution model is $O(\delta^{2N+2})$; it satisfies*

$$\int_\Omega |\tau_N(u,u)| dx \ \leq \delta^{2N+2} ||\triangle^{N+1}(-\delta^2\triangle + 1)^{-(N+1)}u|| ||u||,$$

$$\leq C\delta^{2N+2} ||\triangle^{N+1}u|| ||u||.$$

Proof. By the Cauchy-Schwarz inequality

$$\int_\Omega |\tau_N(u,u)| dx \leq ||u - D_N(\overline{u})|| ||u||$$

$$\leq \delta^{2N+2} ||\triangle^{N+1}(-\delta^2\triangle + 1)^{-(N+1)}u|| ||u|| \ .$$

$\square$

6.4.1 The Case of Homogeneous, Isotropic Turbulence

Precise estimates of time averaged consistency errors can be given following [LL08]. The Leray/Leray-alpha model is the case $N = 0$ in the family of

Leray-deconvolution models so we shall denote the consistency error tensor
of the Leray model by $\tau_0(u, u)$ where

$$\tau_0(u, u) := \overline{u}u - uu,$$

$$\text{where } \overline{u} = (-\delta^2\triangle + 1)^{-1}u .$$

Time averaged values are always well defined and can be estimated in terms
of model parameters and the Reynolds number as in [LL05].

Definition 54. Let $< \cdot >$ denote long time averaging, given by

$$< \phi >:= \lim_{T\to\infty} \sup \frac{1}{T} \int_0^T \phi(t)dt . \qquad (6.21)$$

Let $\varepsilon(u)$ and ε denote respectively the energy dissipation rate of the flow (the
unknown true solution of the Navier–Stokes equations) and its time averaged
value, defined by

$$\varepsilon(u)(t) = \frac{1}{L^3} \int_\Omega \nu|\nabla u|^2 dx, \text{ and}$$

$$\varepsilon =< \varepsilon(u)(t) > .$$

Lemma 55. *If $f \in L^\infty(0, T; H^{-1}(\Omega))$ then $<\varepsilon(u)><\infty$ and $<\varepsilon(w)><\infty$.*

Proof. The Navier–Stokes equations result is standard, e.g., Doering and
Gibbon [DG95]. The result for the model is proven the same way (divide
the energy equality by T and take the limit as $T \to \infty$. Using Gronwall's
inequality, $\frac{1}{T}||w(T)||^2 \to 0$. The remainder follows from the Cauchy-Schwarz
inequality).   ☐

Using the estimate of the regularization's accuracy, a sharp estimate of the
Leray model's consistency error can be given. Naturally, for a general weak
solution of the Navier–Stokes equations, it is not known if the indicated norms
on the RHS are finite at all times.

Proposition 56. *Let*

$$U :=< \frac{1}{L^3} \int_\Omega |u(x, t)|^2 dx >^{\frac{1}{2}}$$

*The consistency error of the Leray model and Leray-alpha model satisfy, for
smooth u,*

$$\int_\Omega |\tau_0(u, u)|dx \le C\delta||u|| \min\{||\nabla u||, \delta||\triangle u||\}.$$

The time averaged consistency error of the Leray model and Leray-alpha model satisfy, for smooth u,

$$< \frac{1}{L^3} \int_\Omega |\tau_0(u,u)| dx >$$

$$\leq C\delta < \frac{1}{L^3} ||u||^2 >^{\frac{1}{2}} \min\{< \frac{1}{L^3} ||\nabla u||^2 >^{\frac{1}{2}}, \delta < \frac{1}{L^3} ||\triangle u||^2 >^{\frac{1}{2}}\}$$

$$< \frac{1}{L^3} \int_\Omega |\tau_0(u,u)| dx > \leq C\frac{\delta}{L} \operatorname{Re} < \varepsilon >^{\frac{1}{2}} .$$

Finally,

$$< \frac{1}{L^3} \int_\Omega |\tau_0(u,u)| dx > \leq C\frac{\delta}{\sqrt{\nu}}(\frac{U^3}{L})^{\frac{1}{2}} U^{\frac{1}{2}} = C\frac{\delta}{L} Re^{\frac{1}{2}} U^2.$$

Proof. By the Cauchy-Schwarz inequality

$$\int_\Omega |\tau_0(u,u)| dx \leq ||u - \overline{u}|| \, ||u||.$$

Both filters satisfy

$$||u - \overline{u}|| \leq C\delta \min\{||\nabla u||, \delta ||\triangle u||\},$$

e.g., for the Leray-alpha model $C = 1$. Integrate this in time, apply the temporal Cauchy-Schwarz inequality, take limits superior:

$$\left\langle \frac{1}{L^3} ||u - \overline{u}||^2 \right\rangle \leq C\delta^2 \left\langle \frac{1}{L^3} ||\nabla u||^2 \right\rangle = C\frac{\delta^2}{\nu} \left\langle \frac{\nu}{L^3} ||\nabla u||^2 \right\rangle = C\frac{\delta^2}{\nu} \langle \varepsilon \rangle$$

$$\square$$

It is useful to estimate the RHS of these bounds in terms of the Reynolds number and (after time averaging) for a general weak solution of the Navier–Stokes equations. We have

$$< \frac{1}{L^3} \int_\Omega |\tau_0(u,u)| dx > \leq C\delta < \frac{1}{L^{\frac{3}{2}}} ||\nabla u|| \frac{1}{L^{\frac{3}{2}}} ||u|| >, \text{ or}$$

$$< \frac{1}{L^3} \int_\Omega |\tau_0(u,u)| dx > \leq C\frac{\delta}{\sqrt{\nu}} < \frac{\nu}{L^3} ||\nabla u||^2 >^{\frac{1}{2}} < \frac{1}{L^3} ||u||^2 >^{\frac{1}{2}} .$$

Rewriting these in terms of the non-dimensionalized quantities gives

$$< \frac{1}{L^3} \int_\Omega |\tau_0(u,u)| dx > \leq C\frac{\delta}{\sqrt{\nu}} \varepsilon^{\frac{1}{2}} U.$$

Now, in turbulent flow typically $\varepsilon \approx \frac{U^3}{L}$ by dimensional analysis, Pope [Po00], and experiment, [S84], [S98]. The upper estimate $\varepsilon \leq C\frac{U^3}{L}$ has also been proven directly from the Navier–Stokes equations. Thus we have

$$< \frac{1}{L^3} \int_\Omega |\tau_0(u,u)| dx > \leq C \frac{\delta}{\sqrt{\nu}} \left(\frac{U^3}{L}\right)^{\frac{1}{2}} U = C \frac{\delta}{L} Re^{\frac{1}{2}} U^2;$$

Related and more detailed estimates can be obtained in the case of homogeneous, isotropic turbulence using the techniques introduced in [LL05].

6.5 Microscales

When the higher order Leray-deconvolution models are used to approximate turbulent flows, an estimate of computational resources required can be obtained by estimating (under the assumptions of isotropy and homogeneity) the model's microscale. This was first performed by Muschinsky [Mus96] for the Smagorinsky model and has been used for other models recently, e.g. [LN06a]. We conclude our study of Leray-deconvolution models and their discretization by summarizing this analysis.

The Reynolds number for Navier–Stokes equations represents the ratio of nonlinearity over the viscous terms. Then, for the Leray-deconvolution models we have

$$Re_{model} \simeq \frac{|D_N \overline{u} \cdot \nabla u|}{|\nu \Delta u|}.$$

The model's Reynolds numbers with respect to the model's largest and persistent scales are thus

$$Re_{model-large} = \frac{UL}{\nu(1 + (\frac{\delta}{L})^2)} \sum_{n=0}^{N} \left(1 - \frac{1}{1 + \frac{\delta^2}{L^2}}\right)^n$$

$$Re_{model-small} = \frac{w_{small} \eta_{model}}{\nu(1 + (\frac{\delta}{\eta_{model}})^2)} \sum_{n=0}^{N} \left(1 - \frac{1}{1 + \frac{\delta^2}{\eta_{model}^2}}\right)^n.$$

where η_{model} is the models' microscale, i.e. the length scale of the models' smallest persistent eddies and w_{small} is the characteristic velocity of the model's smallest persistent eddies.

As in the Navier–Stokes equations, any energy cascade in the Leray-deconvolution models is halted by viscosity grinding down eddies exponentially fast when

$$Re_{model-small} = O(1), \text{ i.e., when}$$

$$\frac{w_{small}\eta_{model}}{\nu(1+(\frac{\delta}{\eta_{model}})^2)} \sum_{n=0}^{N} \left(1 - \frac{1}{1+\frac{\delta^2}{\eta_{model}^2}}\right)^n \simeq 1.$$

Thus,

$$w_{small} \simeq \frac{\nu}{\eta_{model}} \left(\frac{1}{(1+\frac{\delta^2}{\eta_{model}^2})} \sum_{n=0}^{N} \left(1 - \frac{1}{1+\frac{\delta^2}{\eta_{model}^2}}\right)^n\right)^{-1}.$$

The second important equation determining the model's micro-scale comes from statistical equilibrium, i.e., matching energy in to energy out. The rate of energy input to the largest scales is the energy over the associated time scale

$$\frac{E_{model}}{(\frac{L}{U})} = \frac{U^2}{(\frac{L}{U})} = \frac{U^3}{L}.$$

When the model reaches statistical equilibrium, the energy input to the largest scales must match the energy dissipation at the model's micro-scale which scales like $\varepsilon_{small} \simeq \nu(|\nabla w_{small}|^2) \simeq \nu(\frac{w_{small}}{\eta_{model}})^2$. Thus we have

$$\frac{U^3}{L} \simeq \nu(\frac{w_{small}}{\eta_{model}})^2.$$

Inserting the above formula for the micro-eddies characteristic velocity w_{small} gives

$$\frac{U^3}{L} \simeq \frac{\nu^3}{\eta_{model}^4} \left(\frac{1}{(1+\frac{\delta^2}{\eta_{model}^2})} \sum_{n=0}^{N} \left(1 - \frac{1}{1+\frac{\delta^2}{\eta_{model}^2}}\right)^n\right)^{-2} \tag{6.22}$$

Next, the solution to this equation depends on which term in the numerator of the RHS is dominant: 1 or $(\frac{\delta}{\eta_{model}})^2$.

6.5.1 Case 1

when $\delta \ll \eta_{model}$, i.e. $1 + \frac{\delta^2}{\eta_{model}^2} \simeq 1$. Then

$$\eta_{model} \simeq Re^{-\frac{3}{4}}L \ , \quad \text{for } N = 0, 1, 2, \ldots.$$

For case 1 the model predicts the correct microscale, i.e. Kolmogorov microscale since that case occurs when the averaging radius δ is so small that the model is very close to the NSE. However, the latter case is the expected case.

6.5.2 Case 2

when $\delta \gg \eta_{model}$, i.e. $1 + \frac{\delta^2}{\eta_{model}^2} \simeq \frac{\delta^2}{\eta_{model}^2}$. We rewrite equation (6.22).

$$\frac{U^3}{L} \simeq \frac{\nu^3}{\eta_{model}^4} \left(\frac{1}{(1 + \frac{\delta^2}{\eta_{model}^2})} \sum_{n=0}^{N} \left(\frac{\frac{\delta^2}{\eta_{model}^2}}{1 + \frac{\delta^2}{\eta_{model}^2}} \right)^n \right)^{-2}$$

Since $\delta \gg \eta_{model}$ we have

$$\frac{U^3}{L} \simeq \frac{\nu^3}{\eta_{model}^4} \left(\frac{1}{\frac{\delta^2}{\eta_{model}^2}} \sum_{n=0}^{N} \left(\frac{\frac{\delta^2}{\eta_{model}^2}}{\frac{\delta^2}{\eta_{model}^2}} \right)^n \right)^{-2}$$

Therefore,

$$\eta_{model} \simeq Re^{-\frac{3}{4}} L^{1/2} \delta^{1/2} (N+1)^{1/8} \quad \text{for} \quad N = 0, 1, 2, \ldots.$$

The microscale of the Leray-deconvolution models is larger than the Kolmogorov microscale which is $O(Re^{-3/4})$. An interesting result which was observed experimentally is that the models' microscale is affected by the order of the deconvolution operator, meaning that the increase of N gives more truncation of small scales but preserving high accuracy of the models' solution over the large scales.

6.6 Discretization

Work in [LMNR08b] has shown the Leray-deconvolution is able to achieve excellent results in computations. Given inf-sup stable discrete spaces

$$(X_h, Q_h) \subset (H_0^1, L_0^2) \text{ or } (H_\#^1, L_\#^2),$$

let V_h denote the discretely divergence free subspace of X_h. Define the discrete filter by: Given $\phi \in L^2$, $G_h \phi := \overline{\phi}^h$ is the unique solution in X_h to

$$\delta^2 (\nabla \overline{\phi}^h, \nabla v_h) + (\overline{\phi}^h, v_h) = (\phi, v_h) \quad \forall v_h \in X_h. \tag{6.23}$$

For internal flows, recent work in [BR10] shows an incompressible filter can yield improved results:

$$\delta^2(\nabla\overline{\phi}^h, \nabla v_h) + (\overline{\phi}^h, v_h) - (\lambda, \nabla \cdot v_h) = (\phi, v_h) \ \forall v_h \in X_h,$$

$$(\nabla \cdot \overline{\phi}^h, q_h) = 0 \ \forall q_h \in Q_h$$

Discrete approximate deconvolution is defined by the discrete filter as

$$D_N^h = \sum_{n=0}^{N} (I - G_h)^n. \tag{6.24}$$

These definitions provide us with the following Crank-Nicolson FEM for the Leray-deconvolution model: Given w_0, let w_0^h be its L^2 projection into V_h, and for $n = 1, ..., M - 1$,

$$\frac{1}{\Delta t}(w_{n+1}^h - w_n^h, v^h) + b^*(\overline{D_N^h w_{n+1/2}^h}^h, w_{n+1/2}^h, v^h) + \nu(\nabla w_{n+1/2}^h, \nabla v^h)$$

$$= (f_{n+1/2}, v^h) \ \forall \ v^h \in V^h. \tag{6.25}$$

In [LMNR08b], this numerical scheme was shown to be unconditionally stable.

Lemma 57. *Solutions to (6.25) exist and satisfy*

$$\|w_M^h\|^2 + \nu\Delta t \sum_{n=0}^{M-1} \|\nabla w_{n+1/2}^h\|^2 \leq \|w_0^h\|^2 + \frac{\Delta t}{\nu}\|f_{n+1/2}\|_*^2. \tag{6.26}$$

The following convergence estimate is proved in [LMNR08b]:

Theorem 58. *Let (u, p) be a smooth solution to the NSE, and (X_h, Q_h) denote the finite element spaces of (P_k, P_{k-1}) velocities and pressures (respectively) on a locally quasi-uniform finite element mesh and for $k \geq 2$. Assuming the first k partial derivatives of u vanish on the boundary or the BCs are periodic and for small enough Δt,*

$$\|u - w_h\|_{l^\infty(0,M;L^2)} + \left(\nu\Delta t \sum_{n=0}^{M-1} \|\nabla(u(t_{n+1/2}) - w_{n+1/2}^h)\|^2\right)^{1/2}$$

$$\leq C(h^k + \Delta t^2 + \delta^{2N+2}). \tag{6.27}$$

Remark 59. The necessity of the assumption that partial derivatives of the velocity vanishing on the boundary is discussed in the Appendix.

Remark 60. The convergence estimate provides a balancing on the choice of N, k, δ, h, Δt, for optimal convergence. Experience has shown raising N above what is required for optimal convergence (given the other parameters) can still improve error.

Preliminary computations with the scheme (6.25) were done in [LMNR08b, BR10] and gave promising results. Continuing study of this and related methods is ongoing.

6.7 Numerical Experiments with Leray-Deconvolution

We present now three numerical experiments using the scheme (6.25), a convergence rate verification, and 2D and 3D channel flow over a forward-backward facing step. These examples will demonstrate how the use of deconvolution can significantly increase the accuracy. For both examples, we linearize (6.25) by linearly extrapolating the regularized velocity in the nonlinearity. It is shown in [BR10] that this does not alter the unconditional stability or optimal convergence of the algorithm, but it does decouple the conservation system from the regularization.

6.7.1 Convergence Rate Verification

Our first test is to verify the predicted convergence rates proven in Theorem 58 with $N = 0, 1$. We use the Chorin problem for our tests, which is an NSE solution in $\Omega = (0,1) \times (0,1)$ with the form

$$u_1(x,y,t) = -\cos(n\pi x)\sin(n\pi y)e^{-2n^2\pi^2 t/\nu}$$

$$u_2(x,y,t) = \sin(n\pi x)\cos(n\pi y)e^{-2n^2\pi^2 t/\nu}$$

$$p(x,y,t) = -\frac{1}{4}(\cos(2n\pi x) + \cos(2n\pi y))e^{-2n^2\pi^2 t/\nu}$$

For these calculations, we set $\alpha = h$, $\nu = 0.3$, $u_0 = u(0)$, $T = 0.001$, and f from u, p and the NSE, and used (P_3, P_2) Taylor-Hood elements. The increased order of convergence for $N = 1$ can be seen in the $L^2(0, T; H^1(\Omega))$ norm, over the $N = 0$ case. These rates, second order for $N = 0$ and third order for $N = 1$, are given in Table 6.1, agree with Theorem 58, and illustrate the increase in accuracy offered by deconvolution. We note that the small ending time is used because of the 'large' ν, and thus to suppress temporal error arising from fast exponential decay of the solution.

Table 6.1 $L^2(0,T;H^1(\Omega))$ errors and rates found with the incompressible α-filter for experiment 1. A higher convergence rate can be observed for $N = 1$ (RIGHT) versus $N = 0$ (LEFT)

h	Δt	$\|u - u^h_{LD0}\|_{2,1}$	Rate	$\|u - u^h_{LD1}\|_{2,1}$	Rate
1/4	0.001	0.00087455	–	0.00086246	–
1/8	0.001/3	0.0001156	2.919	0.00011018	2.969
1/16	0.001/9	1.5617×10^{-5}	2.888	1.4046×10^{-5}	2.972
1/32	0.001/27	2.2311×10^{-6}	2.807	1.7639×10^{-6}	2.993
1/64	0.001/81	3.7527×10^{-7}	2.572	2.204×10^{-7}	3.001

Table 6.2 $L^2(0,T;H^1(\Omega))$ errors and rates for experiment 1. The regular α-filter (i.e. without enforcing incompressibility) is used

h	Δt	$\|u - u^h_{LD0}\|_{2,1}$	Rate	$|u - u^h_{LD1}|_{2,1}$	Rate
1/4	0.001	0.012751	–	0.012731	–
1/8	0.001/3	0.0017053	2.903	0.0016726	2.928
1/16	0.001/9	0.00024814	2.781	0.00021161	2.983
1/32	0.001/27	4.3274×10^{-5}	2.520	2.6794×10^{-5}	2.981
1/64	0.001/81	7.1642×10^{-6}	2.595	3.6854×10^{-6}	2.862

In order to achieve optimal convergence rates, our experimentation showed the necessity of an incompressible filter, rather than just the usual discrete α-filter. That is, with an incompressible filter and (P_3, P_2) elements, spatial error of $O(h^3)$ is obtained. With the usual α-filter, however, $O(h^3)$ spatial convergence appears to not be achieved. Table 6.2 contains the errors and rates for the schemes' approximation to the same problem, using the usual α-filter. In the $N = 1$ case, the rate of convergence appears to decay away from 3.

6.7.2 Two-Dimensional Channel Flow Over a Step

Our first experiment is for two-dimensional flow over a forward and backward facing step, where the domain Ω is a 40 x 10 channel with a 1 x 1 step five units into the channel at the bottom. No-slip boundary conditions on enforced on the top and bottom boundaries, and parabolic inflow and outflow profiles are enforced, given by $(y(10 - y)/25, 0)^T$. Computations were performed on two mesh levels, with total number of degrees of freedom 5091 and 8927, respectively. The correct physical behavior, as found in [LMNR08b, JL06], is a smooth velocity field that has eddies form and shed behind the step.

We first compute the NSE directly (the same scheme, but no regularization). Computations were made on both mesh levels, using $\Delta t = 0.01$, and $\nu = 1/600$, (P_3, P_2) elements. Comparing to known DNS data in [LMNR08b, JL06], the coarse mesh solution at $T = 40$, shown in Fig. 6.1,

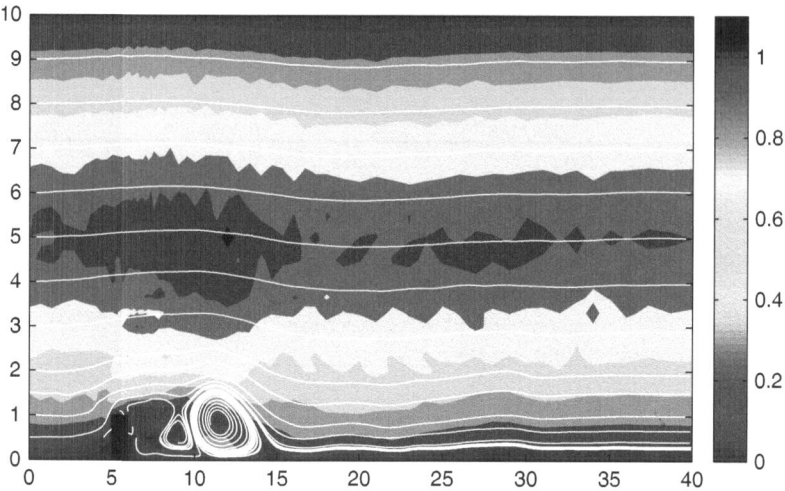

Fig. 6.1 Navier–Stokes Equations on a Coarse Mesh, $T = 40$: Solution incorrect as oscillations are present in speed contours

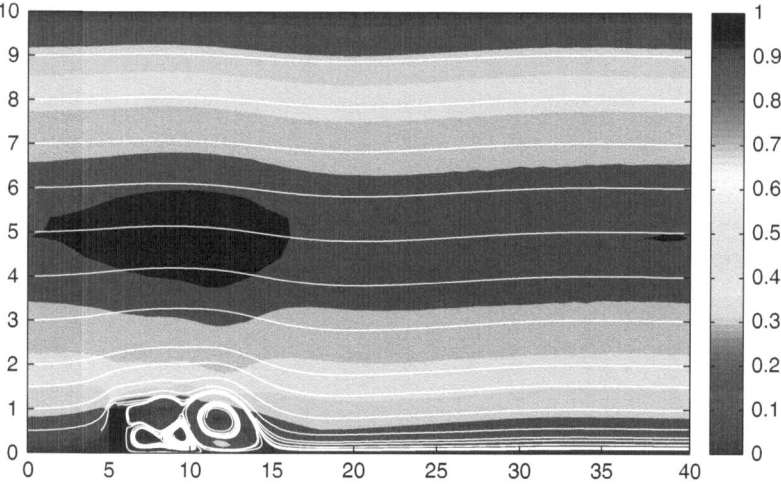

Fig. 6.2 Navier–Stokes Equations on a Fine Mesh, $T = 40$: "Truth" Solution

is incorrect, as it has significant oscillations present and does not completely capture eddy reformation after detachment. The solution from the finer mesh (Fig. 6.2) agrees reasonably well with DNS results.

Since the goal of fluid flow models is to predict the correct solution on coarser meshes than a DNS requires, we test the scheme on the coarse mesh, and obtained the following results using the same parameters as for the coarse

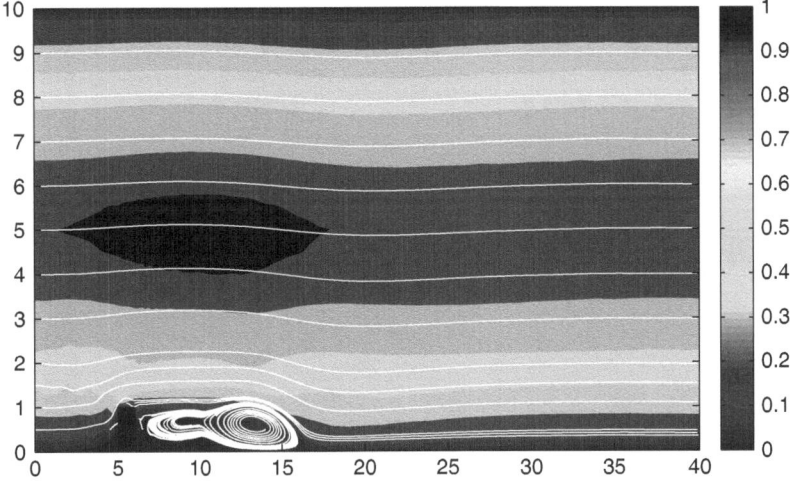

Fig. 6.3 Leray CNLE, N = 0, $T = 40$: Smooth flow, but eddy detachment incorrect

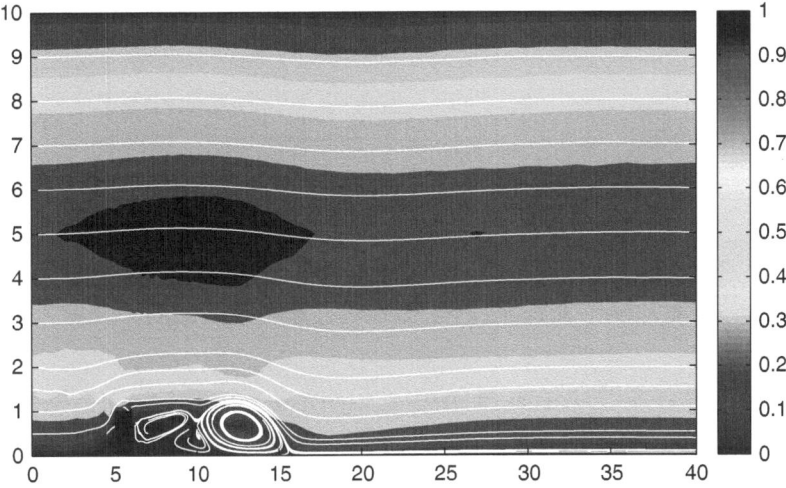

Fig. 6.4 Leray CNLE, N = 1, $T = 40$, $\Delta t = 0.01$: Smooth flow and correct eddy detachment and reformation found

mesh NSE computation and with $\alpha = 1$ (approximately the average element diameter). The $N = 0$ model (i.e. Leray-α) found a smooth flow field, but was unable to correctly predict eddy detachment and reformation (Fig. 6.3). The $N = 1$ model, however, was able to capture the correct behavior (Fig. 6.4). Interestingly, for $N = 1$, we were able to increase the timestep to 0.1 and still get approximately the same $T = 40$ picture (Fig. 6.5).

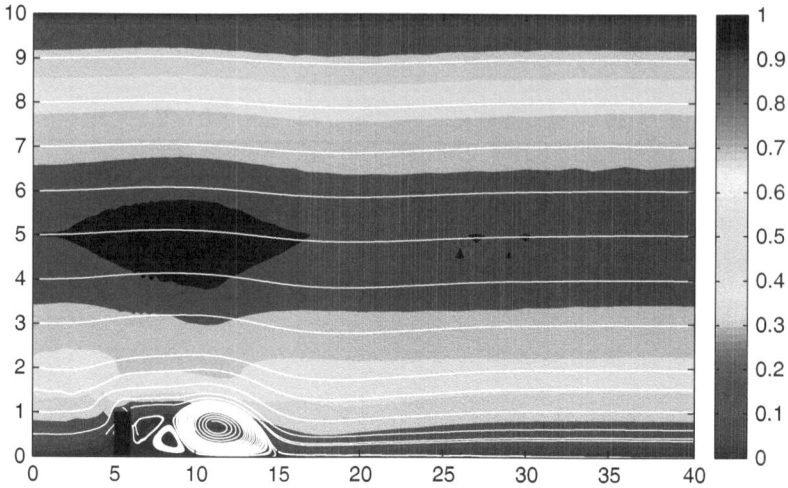

Fig. 6.5 Leray CNLE, N = 1, $\Delta t = 0.1$, $T = 40$

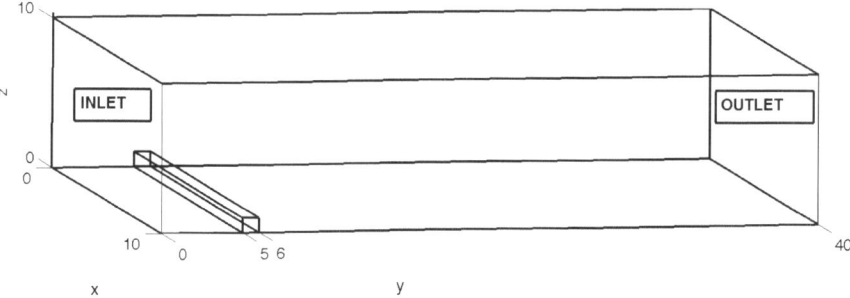

Fig. 6.6 Shown above is domain for the 3D channel flow over a step problem

6.7.3   Three-Dimensional Channel Flow Over a Step

This second experiment is a three-dimensional analog of the two-dimensional step problem above, but now in a 3D channel and with $Re = 200$. A diagram of the flow domain is given in Fig. 6.6; the channel is $10 \times 40 \times 10$ with a $10 \times 1 \times 1$ block on the bottom of the channel, 5 units in from the inlet. No slip boundary conditions are enforced on the channel walls and on the step. The initial condition is that of the $Re = 50$ steady flow on this domain, and an inflow=outflow condition is enforced. We compute with (P_3, P_2^{disc}) Scott-Vogelius elements on a barycenter refined tetrahedralization of the domain, via the method developed in [OR11], and a uniform timestep of $\Delta = 0.025$ is used to compute solutions up to end-time $T = 10$. This flow was studied in [CRW10], and similar to the two-dimensional problem, the correct behavior at $T = 10$ is for an eddy to have detached from behind

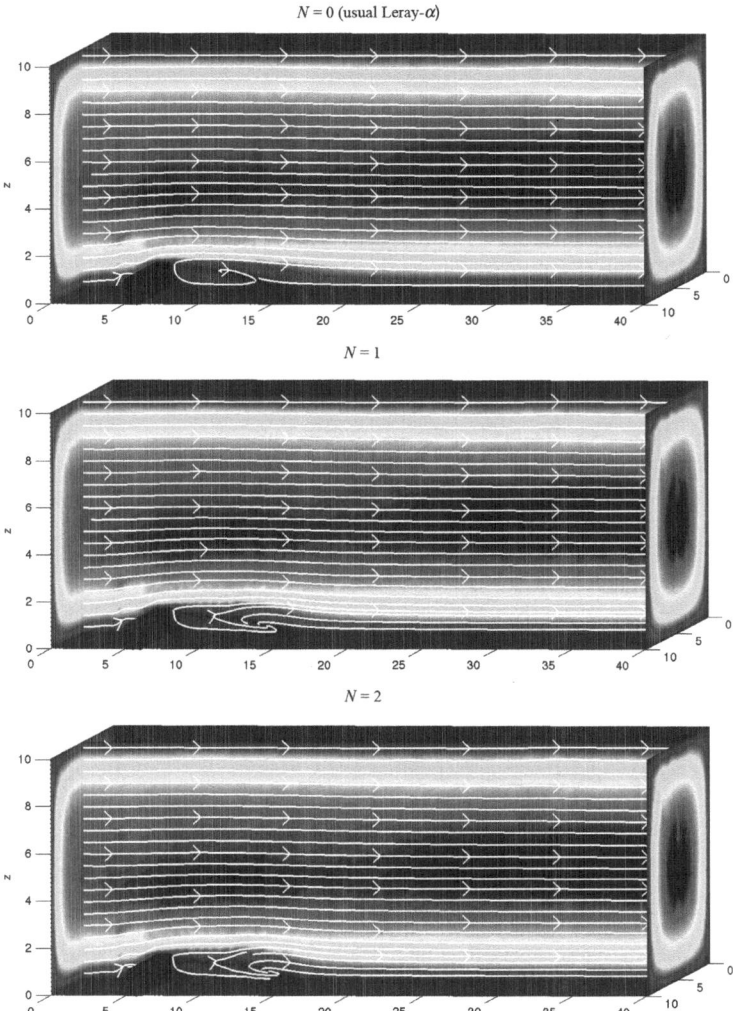

Fig. 6.7 Shown above is velocity streamlines and speed contours on the $x = 5$ sliceplanes of the $T = 10$ velocity solutions for 3D flow over a step, for Leray-deconvolution, with $N = 0$, 1, 2, from top to bottom

the step and moved down the channel, and a new eddy to form. For the direct numerical simulation of this problem that was performed in [CRW10], the total spacial degrees of freedom needed to resolve this flow was 1,282,920. We test our methods on a much coarser mesh that gives 193,596 total degrees of freedom.

Plots of the solutions obtained by the methods for $N = 0$ (usual Leray-α), $N = 1$ and $N = 2$ are shown at $T = 10$ in Fig. 6.7, at the $x = 5$ mid-slice plane, as streamlines over speed contours. All three solutions correctly predict the smooth velocity field, but the detaching of eddies is only correct when deconvolution is used.

6.8   Remarks

This chapter is based on the work in [LMNR08b] and [LL08]; see also [BR10, LMNR09] for further developments. There is considerable analysis of the Leray-alpha model, which is the N=0 case of the Leray deconvolution regularization, e.g., [CHOT05, ILT05, VTC05]. In computational studies of accuracy of the family of models as approximations to flow problems, deconvolution plays an essential role. Thus, extending many of these analytic results from the $N = 0$ case to the important case of higher N (and the entire family) is an important open problem.

Chapter 7
NS-Alpha- and NS-Omega-Deconvolution Regularizations

"... his natural curiosity displayed itself at an early age. ... 'Show me how it doos' is never out of his mouth. He also investigates the hidden course of streams and bell-wires, the way the water gets from the pond through the wall and a pen or small bridge and down a drain ..." from: http://www-history.mcs.st-andrews.ac.uk/Biographies/Maxwell.html, see also: L Campbell and W Garnett, The life of James Clerk Maxwell with selections from his correspondence and occasional writings (London, 1884).

7.1   Integral Invariants of the NSE

Consider the NSE in rotational form:

$$u_t + (\nabla \times u) \times u - \nu \triangle u + \nabla P = f(x,t),$$

$$\nabla \cdot u = 0,$$

$$\text{where}\quad P = p + \frac{1}{2}|u|^2.$$

When the NSE become the Euler equations whose integral invariants are important for understanding the behavior of NSE solutions at high Reynolds numbers. We impose L periodic boundary conditions under the usual zero mean condition

$$u(x + Le_j, t) = u(x,t) \quad j = 1,2,3, \tag{7.1}$$

$$\int_\Omega \phi \, dx = 0 \quad \text{for} \quad \phi = u,\, u_0,\, f,\, P.$$

W.J. Layton and L.G. Rebholz, *Approximate Deconvolution Models of Turbulence*, Lecture Notes in Mathematics 2042, DOI 10.1007/978-3-642-24409-4_7, © Springer-Verlag Berlin Heidelberg 2012

There are four important integral invariants of the Euler equations (with density normalized to $\rho \equiv 1$)

$$\text{Mass flux } M(t) := \frac{1}{|\Omega|} \int_\Omega u(x,t)dx$$

$$\text{Energy (2d\&3d) } E(t) := \frac{1}{|\Omega|} \int_\Omega \frac{1}{2} |u(x,t)|^2 dx,$$

$$\text{Helicity (3d) } H(t) := \frac{1}{|\Omega|} \int_\Omega \nabla \times u(x,t) \cdot u(x,t)dx,$$

$$\text{Enstrophy (2d) } Ens(t) := \frac{1}{2|\Omega|} \int_\Omega |\nabla \times u(x,t)|^2 dx.$$

There are two additional important integral quantities for the Navier–Stokes equations in $3d$ and one in $2d$

$$\text{Energy dissipation rate } \varepsilon(t) := \frac{1}{|\Omega|} \int_\Omega \nu |\nabla u(x,t)|^2 dx,$$

$$\text{Helicity dissipation rate } \gamma(t) := \frac{2}{|\Omega|} \int_\Omega \nu \nabla \times \omega(x,t) \cdot \omega(x,t)dx,$$

$$\text{Enstrophy dissipation rate (in } 2d) := \frac{1}{|\Omega|} \int_\Omega \nu |\nabla \times \omega(x,t)|^2 dx.$$

The interplay between energy and helicity is thought to organize coherent structures in fluid motion. Thus, good regularizations should also conserve both in $3d$. The 2d Euler equations conserve enstrophy exactly. This exact conservation is why large structures tend to form in $2d$ flows (and approximately in large thin layers of 3d flows) whereas in fully 3d flows large structures are relentlessly ground down by nonlinearity. Thus, a regularization that is to be used to simulate the motion of large thin layers of fluids should also conserve some form of model enstrophy exactly.

We shall see that, while the Leray regularization only conserves kinetic energy (in the appropriate context), both the NS-alpha and NS-omega regularizations, developed in this chapter, exactly conserve both energy and helicity. The NS-α regularization exactly conserves (under periodic boundaries and no viscous or external forces) the two key integral invariants for three dimensional fluid flow: an approximate model energy, $\frac{1}{2}(w, \overline{w})$, and helicity, $(w, \nabla \times w)$. The NS-ω regularization exactly conserves (again under periodic boundaries and no viscous or external forces) the exact kinetic energy, $\frac{1}{2}\|w\|^2$, and an approximate model helicity, $(w, \nabla \times \overline{w})$.

7.2   The NS-Alpha Regularization

The NS-α model is a recently developed regularization of the NSE with desirable mathematical properties, e.g., [FHT01, FHT02, GOP03]. It is given by

$$w_t - \overline{w} \times (\nabla \times w) + \nabla Q - \nu \Delta w = f, \text{ in } \Omega \times (0, T], \qquad (7.2)$$

$$\nabla \cdot \overline{w} = 0, \text{ in } \Omega \times (0, T], \qquad (7.3)$$

$$-\delta^2 \Delta \overline{w} + \overline{w} = w, \text{ in } \Omega \times (0, T], \qquad (7.4)$$

$$w(x, 0) = u_0(x), \text{ in } \Omega. \qquad (7.5)$$

We now want to consider the NS-alpha regularization for practical simulations. Thus, we must move away from periodic boundary conditions. There are very many approaches to physical and numerical BCs in CFD. It does seem plausible however that the place to begin is with no slip BCs. Thus we shall impose

$$w = 0, \text{ on } \partial\Omega.$$

Under no slip BCs, the question of what is the right filter again arises. In the non periodic case the simple differential filter does not preserve incompressibility. Thus, one natural idea is to pay the extra computational price and filter by solving a shifted Stokes problem instead of a Poisson problem. This means defining the filter by

$$-\delta^2 \triangle \overline{w} + \overline{w} + \nabla \lambda = w, \text{ and } \nabla \cdot \overline{w} = 0 \text{ in } \Omega \times (0, T],$$

$$\overline{w} = w(= 0) \text{ on } \partial\Omega,$$

and adding the constraint $\nabla \cdot w = 0$ to the system (7.2)–(7.5). Computational practice and the refinements of the models that result will determine when and how the Stokes filter can be replaced by the less expensive differential filter.

7.2.1   The Periodic Case

We first review the building blocks of its theory in the case of periodic boundary conditions. Thus, to begin, consider the NS-alpha model written in rotation form. Since it has been intensively studied over some years already, there has been time to reflect on the essential features of the model and distill the original complex and brilliant arguments. Thus, (as happens with many successful theories), the theory of the model is simpler in retrospect after this distillation. In the periodic case (i.e., when $\Omega = (0, L)^3$), the filter used is simply

$$\overline{w} := (-\delta^2 \triangle + 1)^{-1} w, \tag{7.6}$$

subject to L-periodic boundary conditions. Thus,[1] we can multiply the model by $\overline{w}$ and integrate over the flow domain. This gives

$$(w_t, \overline{w}) - \nu(\triangle w, \overline{w}) = (f, \overline{w}), \tag{7.7}$$

since the nonlinearity obviously vanishes when its dot product is taken with both $\overline{w}$ and $\nabla \times w$. The operator $(-\alpha^2 \triangle + 1)^{-1}$ is SPD and

$$w \to (w, (-\delta^2 \triangle + 1)^{-1})^{\frac{1}{2}} = (w, \overline{w})^{\frac{1}{2}}$$

is a norm which is equivalent for fixed $\delta > 0$ to the $H^{-1}(\Omega)$ norm (denoted $||\cdot||_{-1}$ herein). Similarly,

$$w \to (\triangle w, \overline{w})^{\frac{1}{2}}$$

is a norm equivalent to the $L^2(\Omega)$ norm (denoted $||\cdot||$ herein) for fixed $\delta > 0$.

The energy equality and norm equivalence (noted above) implies that

$$w \in L^\infty(0, T; H^{-1}(\Omega)) \cap L^2(0, T; L^2(\Omega)), \tag{7.8}$$

which is the first á priori bound on the solution of the model. This bound is insufficient for an existence theory of weak solutions. However, the averaging in the model's definition makes the nonlinearity smoother (less nonlinear in some sense) and allows a bootstrap argument to proceed, beginning with the first á priori bound, that shows the model's solutions are regular enough (for periodic boundary conditions and $\delta > 0$) for unique strong solutions to exist. We review this argument now.

In the next step of the bootstrap argument, take the inner product of the model with w. This gives

$$(w_t, w) + \nu(\nabla w, \nabla w) = (f, w) + (\overline{w} \times (\nabla \times w), w).$$

Standard inequalities on the RHS imply

$$\frac{1}{2} \frac{d}{dt} ||w||^2 + \frac{\nu}{2} ||\nabla w||^2 \le \frac{1}{2\nu} ||f||_{-1}^2 + C ||\nabla \overline{w}|| ||\nabla w||^{\frac{3}{2}} ||w||^{\frac{1}{2}}.$$

---

[1] We proceed formally in the derivation of the required á priori estimates. As usual, this is a mathematical shorthand for deriving the same estimates for Galerkin approximations in spaces of truncated Stokes eigenfunctions and taking the limit as $N \to \infty$.

Now by the first á priori estimate

$$||\nabla\overline{w}|| \le C(\delta)||w||_{-1} \in L^\infty(0,T).$$

Thus $ess - sup_{(0,T)}||\nabla\overline{w}|| \le C(data, \delta)$ and we have

$$\frac{1}{2}\frac{d}{dt}||w||^2 + \frac{\nu}{2}||\nabla w||^2 \le C(data, \delta) + C(data, \delta)||\nabla w||^{\frac{3}{2}}||w||^{\frac{1}{2}}$$

$$\le C(data, \delta) + \frac{\nu}{4}||\nabla w||^2 + C(\nu, f, w(0), \delta, \Omega)||w||^2.$$

This implies the improved á priori bound

$$w \in L^\infty(0, T; L^2(\Omega)) \cap L^2(0, T; H^1(\Omega)) \text{ and thus}$$

$$\overline{w} \in L^\infty(0, T; H^2(\Omega)) \cap L^2(0, T; H^3(\Omega)).$$

The simple bootstrap argument can now be repeated due to the smoothing in the averaging present in the nonlinear term. Indeed, if A =Stokes Operator, taking the inner product of the model with Aw and repeating the above arguments sketched above gives (since $d \le 3$)

$$\frac{1}{2}\frac{d}{dt}||\nabla w||^2 + \frac{\nu}{2}||\triangle w||^2 \le C(data) + C||\overline{w}||_{L^\infty}||\nabla w||||\triangle w||$$

$$\le C(data) + C||\overline{w}||_{L^\infty(H^2)}||\nabla w||||\triangle w||$$

$$\le C(data) + \frac{\nu}{4}||\triangle w||^2 + C(data)||\nabla w||^2.$$

This implies

$$w \in L^\infty(0, T; H^1(\Omega)) \cap L^2(0, T; H^2(\Omega)) \text{ and thus}$$

$$\overline{w} \in L^\infty(0, T; H^3(\Omega)) \cap L^2(0, T; H^4(\Omega)).$$

This regularity is far more than needed to prove existence and uniqueness of strong solutions of the model and it can be continued even further, if needed.

The limiting behavior of solutions to the model as $\delta \to 0$ has also been studied [FHT02], with results similar to that of Leray-α. Thus the fundamental existence theory of NS-α can be summarized as follows. (There are many more results than are summarized below.)

Theorem 61. *([FHT01, FHT02]) Under periodic boundary conditions and using the differential filter,*

A unique strong solution to NS-α exists.

As the filtering radius $\delta_j \to 0$, solutions of NS-α converge to a weak NSE solution, modulo a subsequence.

Solutions to NS-α conserve a model energy, helicity, and 2d enstrophy, [RS10], under periodic BCs and in the absence of external and viscous forces.

Turbulence phenomenology applied to NS-alpha also predicts that up to a filtering radius dependent length-scale, NS-α cascades energy through the inertial range just as the NSE: $E_{NS\alpha}(k) \sim \epsilon_{NS\alpha}^{1/3} k^{-5/3}$. After this length-scale, energy is cascaded at a faster rate [FHT01].

7.2.2   Discretizations of the NS-alpha Regularization

Consider the above NS-α model. Since the averaged term $\overline{u_{n+1}^h}^h$ is nonlocal, this must be implemented as a coupled system for $(u_{n+1}^h, \phi_{n+1}^h, p_{n+1}^h)$, where $\phi_{n+1}^h = \overline{u_{n+1}^h}^h$. Experience has shown that the schemes that work best are of the following form: $\forall (v^h, \chi^h, q^h, r^h) \in (X^h, X^h, Q^h, Q^h)$,

$$\frac{1}{\Delta t}(u_{n+1}^h - u_n^h, v^h) + ((\nabla \times u_{n+\frac{1}{2}}^h) \times \frac{1}{2}(\phi_{n+1}^h + \overline{u_n^h}^h), v^h)$$

$$-(P_{n+\frac{1}{2}}^h, \nabla \cdot v^h) + \nu(\nabla u_{n+\frac{1}{2}}^h, \nabla v^h) = (f_{n+\frac{1}{2}}, v^h),$$

$$\delta^2(\nabla \phi_{n+1}^h, \nabla \chi^h) + (\phi_{n+1}^h, \chi^h) - (\lambda_{n+1}^h, \nabla \cdot \chi^h) = (u_{n+1}^h, \chi^h),$$

$$(\nabla \cdot \phi_{n+1}^h, q^h) = 0,$$

$$(\nabla \cdot u_{n+1}^h, r^h) = 0.$$

Since (X^h, Q^h) satisfies the LBB^h condition, this is equivalent to, $\forall (v^h, \chi^h) \in (V^h, V^h)$,

$$\frac{1}{\Delta t}(u_{n+1}^h - u_n^h, v^h) + b_\alpha(u_{n+1/2}^h, \frac{1}{2}\phi_{n+1}^h + \frac{1}{2}\overline{u_n^h}^h, v^h)$$

$$+\nu(\nabla u_{n+1/2}^h, \nabla v^h) = (f_{n+1/2}, v^h)$$

$$\delta^2(\nabla \phi_{n+1}^h, \nabla \chi^h) + (\phi_{n+1}^h, \chi^h) = (u_{n+1}^h, \chi^h).$$

Concerning stability, the following has been proven [Con10, RS10]:

Theorem 62. *The fully coupled and nonlinear CN method above is stable for NS-α. For NS-α it is stable with respect to a modified kinetic energy with a modified energy dissipation, given by*

$$KE_\alpha(u) := \frac{1}{2}\|\overline{u}^h\|^2 + \frac{\delta^2}{2}\|\nabla \overline{u}^h\|^2, \quad \varepsilon_\alpha(u) := \nu\|\nabla \overline{u}^h\|^2 + \nu\delta^2\|\triangle \overline{u}^h\|^2.$$

This is second order accurate for NS-α and nonlinearly, unconditionally stable. However, in finite element spaces with N_V velocity degrees of freedom and N_P pressure degrees of freedom, *the method leads to a large, nonlinear system at each time step with $2(N_V + N_P)$ total degrees of freedom.* There does not appear to be any method for NS-α which preserves these attractive properties and has significantly less complexity. Note that once deconvolution is added to the system, the complexity gets even worse. This is the first computational challenge if solving the NS-alpha model: many extra variables must be introduced when using implicit methods. To avoid this, the only alternatives are (a) use a different regularization such as Leray or NS-omega, and (b) use explicit methods for the NS-alpha nonlinearity.

Define

$$b_{rot}(u, w, v) := \int_\Omega (\nabla \times u) \times w \cdot v dx.$$

The most commonly used IMEX method (IMEX comes from IMplicit-EXplicit combination) in CN-AB2 which consists of Crank-Nicolson discretizations of the Stokes terms and second order Adams-Bashforth discretization of the nonlinear terms: $\forall (v^h, \chi^h) \in (V^h, V^h)$,

$$\frac{1}{\Delta t}(w_{n+1}^h - w_n^h, v^h) + \nu(\nabla w_{n+\frac{1}{2}}^h, \nabla v^h)$$

$$+\{\frac{3}{2}b_{rot}(w_n^h, \overline{w_n^h}, v^h) - \frac{1}{2}b_{rot}(w_{n-1}^h, \overline{w_{n-1}^h}, v^h)\} = (f_{n+\frac{1}{2}}, v^h),$$

$$\delta^2(\nabla \phi_{n+1}^h, \nabla \chi^h) + (\phi_{n+1}^h, \chi^h) = (w_{n+1}^h, \chi^h).$$

The main difficulty of IMEX methods is that with highly refined spacial meshes (as in boundary layers) the associated necessary condition for stability

$$\triangle t \leq C \min\{\triangle x : \text{ all local mesh widths } \triangle x\}$$

can be difficult to satisfy.

Similar to the Leray model, NS-α can have at best $O(\alpha^2)$ consistency error. Also similar to the Leray model, this can be improved by approximately deconvolving the filtered terms. However, each order of deconvolution couples another implicit, shifted Stokes problem to the system. Thus this approach to improving accuracy makes an already complex system even worse!

7.3 The NS-Omega Regularization

"However, a dubious increase of understanding was forthcoming at the cost of a considerable increase in mathematical complexity."

Joseph Smagorinsky in: Frontiers of Numerical Mathematics, (R.E. Langer, editor), Univ. of Wisconsin Press, Madison, 1060.

The NS-ω regularization is derived from the rotational form of the NSE by filtering the vorticity term in the nonlinearity (for reasons explained below).

It was first proposed to include a time relaxation term to truncate scales and (using the algorithms considered below) it easily accommodates increasing accuracy with deconvolution operators. The full NS-omega deconvolution model is

$$w_t - w \times (\nabla \times D(\overline{w})) + \nabla Q - \nu \Delta w + \chi(w - D(\overline{w})) = f, \qquad (7.9)$$

$$\nabla \cdot w = 0, \qquad (7.10)$$

$$\overline{w} = G(w). \qquad (7.11)$$

We now consider the NS-omega regularization for CFD simulations. Thus we begin with no slip BCs:

$$w = 0, \text{ on } \partial\Omega.$$

Under no slip BCs, the question of what is the right filter again arises. In the non periodic case the simple differential filter does not preserve incompressibility. Thus, one natural idea is to pay the extra computational price and filter by solving a shifted Stokes problem instead of a Poisson problem. This means defining G by

$$-\delta^2 \triangle \overline{w} + \overline{w} + \nabla \lambda = w, \text{ and } \nabla \cdot \overline{w} = 0 \text{ in } \Omega \times (0, T],$$

$$\overline{w} = w(= 0) \text{ on } \partial\Omega.$$

The simplest NS-omega model is given by (taking $D = I$ and $\chi = 0$)

$$w_t - w \times (\nabla \times \overline{w}) + \nabla Q - \nu \Delta w = f, \qquad (7.12)$$

$$\nabla \cdot w = 0, \qquad (7.13)$$

$$\overline{w} = G(w). \qquad (7.14)$$

Variants of Crank-Nicolson (CN) finite element method (FEM) algorithms have been studied for both NS-ω and NS-α as well as a second order, unconditionally stable, linearly implicit variant (CNLE) of CN for NS-ω.

Proposition 63. *In the absence of viscosity and external forces, and for periodic boundary conditions, the NS-omega model conserves energy and a model helicity.*

Proof. The proof for each result begins by setting $\nu = f = 0$. For energy, multiply by u and integrate over Ω. The nonlinearity and pressure term vanish. This leaves only $\frac{1}{2}\frac{d}{dt}||w||^2 = 0$. Integrating over time gives the result.

The proof for helicity requires multiplying by $\nabla \times \overline{w}$ and integrating over Ω. This leaves $(w_t, \nabla \times \overline{w}) = 0$. Integrating over time and using the fact that differential operators (including the filter) commute under periodic boundary conditions completes the proof. ☐

Many models and regularizations do not exactly conserve (any form of) helicity, [Reb07], and this clearly impacts the reliability of their predictions in highly rotational flows and over longer time intervals. The nonlinearity of NS-ω acts in manner physically consistent with true fluid flow in that it neither creates nor dissipates either energy or helicity. In homogeneous, isotropic turbulence, energy and model helicity conservation suggests that both energy and model helicity in NS-ω cascade from large (input) scales to small scales where they are dissipated strongly by viscosity ([LST08], and the helicity case follows [LMNR09]). It is interesting to note that the three dimensional integral invariants for NS-ω are, in some sense, reversed from those of NS-α. The NS-α model conserves a weaker model energy, $\frac{1}{2}(w, \overline{w})$, than NS-$\omega$, and the usual helicity $(w, \nabla \times w)$.

7.3.1 Motivation for NS-ω: The Challenges of Time Discretization

In the Method, Archimedes described the way in which he discovered many of his geometrical results:
" ... certain things first became clear to me by a mechanical method, although they had to be proved by geometry afterwards because their investigation by the said method did not furnish an actual proof. But it is of course easier, when we have previously acquired, by the method, some knowledge of the questions, to supply the proof than it is to find it without any previous knowledge."–from [Arc02]

Our motivation for the modification from alpha to omega is the search for efficient, unconditionally stable and (at least second order) accurate methods for the simulation of under-resolved flows. The structure of the nonlinearity in NS-ω admits simple methods which are nonlinearly, unconditionally stable, second order accurate, *linearly implicit* (only 1 linear system per time step) and require only $N_V + N_P$ total degrees of freedom. The following attractive variant CNLE (Crank-Nicolson with linear extrapolation, known for the NSE at least since Baker's 1976 paper [B76]) is one possibility. Let $w^{n+1/2} := (w^{n+1} + w^n)/2$ and $U^n := \frac{3}{2}w^n - \frac{1}{2}w^{n-1}$. Then the calculation of $\overline{U^n}$ and $D(\overline{U^n})$ are an explicit and uncoupled calculation of filtering and deconvolution of a known (from previous time values) function. Thus, the following requires no extra storage and only minimal extra operations over the same method for the NSE:

$$\frac{w^{n+1} - w^n}{\triangle t} - \nabla \times D(\overline{U^n}) \times w^{n+\frac{1}{2}} - \nu \triangle w^{n+\frac{1}{2}}$$

$$+ \nabla P^{n+\frac{1}{2}} + \chi(w^{n+1/2} - D(\overline{w^n})) = f^{n+\frac{1}{2}}$$

$$\nabla \cdot w^{n+\frac{1}{2}} = 0 \ .$$

In its full notational glory, this represents, $\forall (v^h, q^h) \in (X^h, Q^h)$,

$$\frac{1}{\Delta t}(w_{n+1}^h - w_n^h, v^h) + b_{rot}(D(\overline{U^n}), \frac{1}{2}w_{n+1/2}^h, v^h)$$

$$-(P_{n+1/2}^h, \nabla \cdot v^h) + \nu(\nabla w_{n+1/2}^h, \nabla v^h) = (f_{n+1/2}, v^h),$$

$$(\nabla \cdot w_{n+1/2}^h, q^h) = 0.$$

Unconditional stability has been proven (take the inner product with $w^{n+\frac{1}{2}}$). It is clearly second order accurate and linearly implicit (since $U^n := \frac{3}{2}w^n - \frac{1}{2}w^{n-1}$ is known from previous time levels). Further, since $U^n := \frac{3}{2}w^n - \frac{1}{2}w^{n-1}$ is known, its average can be directly computed, uncoupled from the linear equations for advancing in time. We also can consider the Crank Nicholson quadratic extrapolation (CNQE), where U^n is extrapolated quadratically as $U^n := \frac{15}{8}w^n - \frac{10}{8}w^{n-1} + \frac{3}{8}w^{n-2}$.

The deconvolution operator involves repeated filtering. Efficiency is preserved if, as here, it acts on a velocity known from previous time levels. The full CN and the CNLE method both are unconditionally stable, well defined and optimally convergent to solutions of the NSE.

Proposition 64. *Both the CN and the CNLE method are unconditionally stable for NS-ω:*

$$\frac{1}{2}\|w_N^h\|^2 + \Delta t \sum_{n=0}^{N-1} \frac{\nu}{2}\|\nabla w_{n+1/2}^h\|^2 \leq \frac{1}{2}\|w_0^h\|^2 + \Delta t \sum_{n=0}^{N-1} \frac{1}{2\nu}\|f_{n+1/2}\|_*^2,$$

Proof. To obtain stability set $v^h = w_{n+1/2}^h$ in the discrete NS-omega equations

$$\frac{1}{2\Delta t}(\|w_{n+1}^h\|^2 - \|w_n^h\|^2) + \nu\|\nabla w_{n+1/2}^h\|^2 \leq \frac{1}{2\nu}\|f_{n+1/2}\|_*^2 + \frac{\nu}{2}\|\nabla w_{n+1/2}^h\|^2$$

i.e.,

$$\frac{1}{\Delta t}(\|w_{n+1}^h\|^2 - \|w_n^h\|^2) + \nu\|\nabla w_{n+1/2}^h\|^2 \leq \frac{1}{\nu}\|f_{n+1/2}\|_*^2,$$

Summing from $n = 0 \ldots N - 1$ gives the desired result. $\square$

7.4   Computational Problems with Rotation Form

"However, for long period integrations, even slow degeneracies may menace the outcome of the results." - J. Smagorinsky, page 118 in: Frontiers of Numerical Mathematics (ed: R.E. Langer, U Wisconsin press, 1960).

The second fundamental difficulty common to both the NS-α and NS-ω is associated with use of the rotational form of the NSE nonlinearity in both (and not with any modeling that has occurred). The nonlinearity in the Navier–Stokes equations and associated regularizations can be written in several ways, which, while equivalent for the continuous NSE, lead to discretizations with different algorithmic costs, conserved quantities, and approximation accuracy, e.g., Gresho and Sani [GS98] and Gunzburger [G89]. These forms include the convective form, the skew symmetric form and the rotation form, given respectively by

$$u \cdot \nabla u, \qquad u \cdot \nabla u + \frac{1}{2}(\text{div}\, u)u, \qquad (\nabla \times u) \times u.$$

The algorithmic advantages and superior conservation properties of the rotation form have led to it being a very common choice for turbulent flow simulations, see, e.g., Chap. 7. in [CHQZ88] and [MM87]. We consider FEM (or other variational) discretizations of the rotation form of the NSE, given by

$$u_t - u \times \boldsymbol{\omega} + \nabla P - \nu \Delta u = f, \tag{7.15}$$

$$\text{div}\, u = 0, \tag{7.16}$$

These are related to the usual NSE form by

$$P = p + \frac{1}{2}|u|^2 \quad , \quad \boldsymbol{\omega} = \text{curl}\, u.$$

It is known from Horiuti [Horiuti87, Horiuti98] and Zang [Z91], see also experiments from [LMNOR09, O02] for the FEM case, that the rotation form can lead to a less accurate approximate solution when discretized by commonly used numerical methods. This loss of accuracy is due to a combination of:

1. The Bernoulli or dynamic pressure $P = p + \frac{1}{2}|u|^2$ is generically much more complex than the pressure p.
2. Meshes upon which p is fully resolved are typically under resolved for P.
3. As the Reynolds number increases, the discrete momentum equation with either form of the nonlinearity magnifies the pressure error's effect upon the velocity error.

In finite element methods (FEM) the inf-sup condition for stability of the pressure places a strong condition linking velocity and pressure degrees of freedom. This condition, while quite technical when precisely stated, roughly implies that for lower order approximations the pressure degrees of freedom should correspond to the velocity degrees of freedom on a mesh one step coarser than the velocity mesh, while for higher order finite elements the polynomial degree of pressure approximations is less than the polynomial degree of velocity approximations. Thus, in either case for velocities u and Bernoulli pressures P with the same complex structures, as the mesh is refined the velocity will be often fully resolved before the Bernoulli pressure is well-resolved.[2]

Example 65. As a simple example, consider Poiseuille flow. In $\Omega = (0,4) \times (0,1)$, take a parabolic inflow $v(x,t) = 0$ and $u(x,y,t) = \frac{1}{2\nu}y(1-y)$ (at $x = 0$). No-slip boundary conditions are given at the top and bottom, and the do-nothing boundary condition is prescribed at the outflow. The exact solution is well known to be

$$v(x,y) = 0, u(x,y) = \frac{1}{2\nu}y(1-y), p(x,y) = -x + 4,$$

provided we take it as our initial condition. If the velocity is discretized with the Taylor-Hood element (quadratic velocities and linear pressures) $u = (u, v)$ and p are in the finite element spaces so that we expect that discretization of the convective and skew symmetric form of the NSE will have very small errors (comparable to the errors from numerical integration and solution of the linear and nonlinear systems arising). On the other hand, if the rotation form is used the exact solution is

$$v(x,y) = 0,$$

$$u(x,y) = \frac{1}{2\nu}y(1-y),$$

$$P(x,y) = p(x,y) + \frac{1}{8\nu^2}y^2(1-y)^2,$$

is not in the pressure finite element space. Thus, in the rotation form there will be discretization errors in P that influence as well the velocity error through the discrete momentum equation, since

$$P \notin Q^h \qquad \text{and} \qquad \|P\| = O(\nu^{-2}). \tag{7.17}$$

The pressure error of the rotation form blows up (because P does) as $Re \to \infty$. This Bernoulli pressure error then influences the velocity error which increases. We shall now study how much influence a large pressure error has on the velocity error and how to reduce its effect with grad div stabilization. This requires some preliminary information about the Stokes problem.

---

[2]When an artificial problem, constructed so the pressure and Bernoulli pressure reverse complexity, is solved the observed error behavior is reversed: the convective form has much greater error than the rotation form.

Even for a simple Prandtl-type, laminar boundary layer, the pressure p will be approximately constant in the near wall region while the Bernoulli pressure $P = p + (1/2)|u|^2$ will share the $O(Re^{-1/2})$ boundary layer of the velocity field. Point 3 is possibly related to aliasing errors; interestingly, the aliasing error in using different forms of the nonlinearity is governed by the resolution of the (Bernoulli or kinematic) pressure. Our suggestion of a "fix" of using $grad-div$ stabilization works in our tests because it addresses point 3 without requiring extra resolution [LMNOR09]. Another effective alternative is to use pointwise divergence-free elements, such as Scott-Vogelius elements. With these elements, rotational form is equivalent to convective form for the NSE [BLCR10], and eliminates the negative scaling effect of the pressure on the velocity error in rotational form models [MNOR11].

7.5 Numerical Experiments with NS-α

We now present two numerical experiments for NS-α with deconvolution following [MNOR11]; these are the same benchmark problems as computed for the Leray model in the previous chapter.

7.5.1 Two-Dimensional Flow Over a Step

We compute with the exact same input parameters and coarse mesh as for the Leray experiment, but now using NS-α. Additionally, we add grad-div stabilization to the method (with parameter 1) for the reasons discussed in the previous section (without this stabilization, the method fails miserably). Solutions for $N = 0$ and $N = 1$ are shown in Figs. 7.1 and 7.2, respectively. Both show a smooth velocity field, but we see a better prediction of eddy separation for $N = 1$.

7.5.2 Three-Dimensional Flow Over a Step

We repeat here the three-dimensional experiment performed for Leray in the previous chapter, but now using NS-α. We linearize (and decouple) the scheme by linearizing the regularized term. Unlike in the case of Leray, unconditional stability is lost, however for this experiment, the timestep is small enough to ensure stability. The solutions are shown for $N = 0$ (usual NS-α), $N = 1$ and $N = 2$ at $T = 10$ in Fig. 7.3, at the $x = 5$ mid-slice plane, as streamlines over speed contours. All three solutions correctly predict the smooth velocity field, but the detaching of eddies is never correctly predicted

$N = 0$ (usual NS-α)

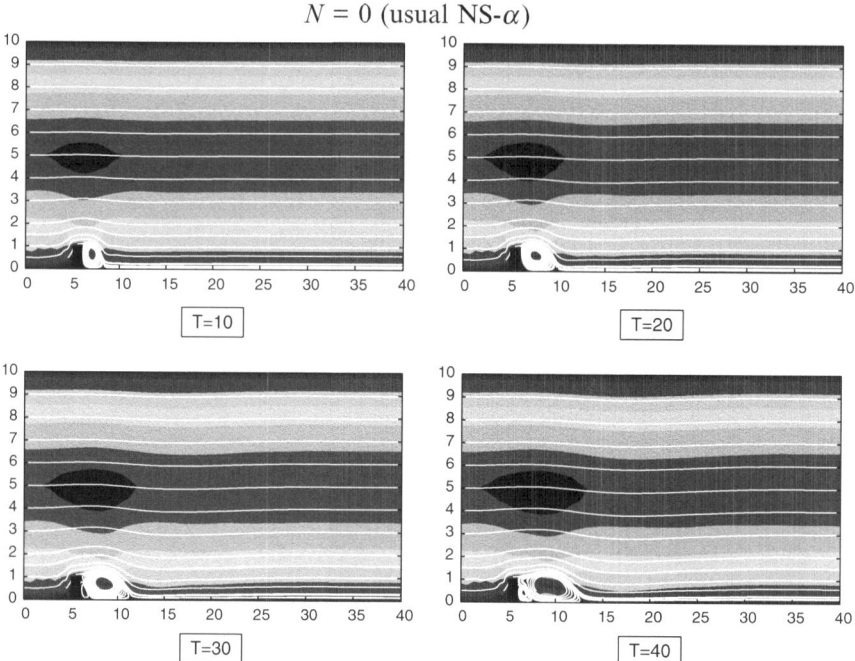

Fig. 7.1 Shown above is velocity streamlines and speed contours of velocity solutions for 2D flow over a step, for NS-α ($N = 0$)

by any of them (as is done by Leray), although there is some improvement with increasing the order of deconvolution.

7.6   Model Synthesis

Reduced models of fluid motion exist as intermediate steps in under-resolved flow simulations. Thus, within any approach to flow modeling the pairing of efficient algorithms with interesting models must be considered as part of the solution process going from fluid phenomena to numerical simulation. Considering possible combinations leads to interesting developments in continuum models as well as algorithms. The form of the regularizations studied, their integral invariants, and development from the deconvolution and NS-α circle of ideas suggest some possibilities for future progress. Three natural ones are synthesis of NS alpha and NS-ω models, higher accuracy in modeling through deconvolution and enhanced scale truncation through combination with VMM / time relaxation ideas.

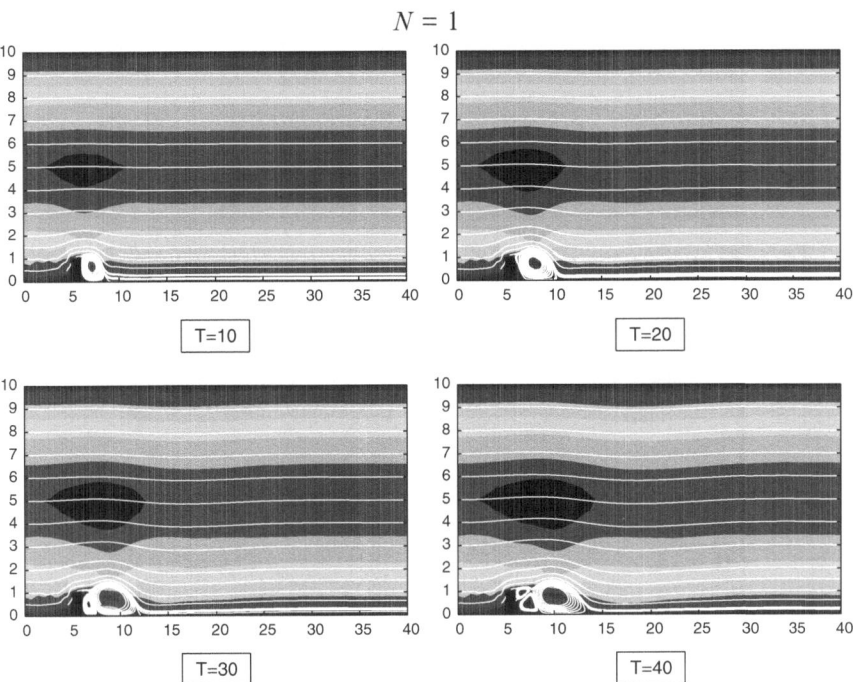

Fig. 7.2 Shown above is velocity streamlines and speed contours of velocity solutions for 2D flow over a step, for NS-α-deconvolution, $N = 1$

7.6.1 Synthesis of NS-α and ω Models

Although our intuition is the contrary, further study could indicate that they perform well in different flow regions. If this is the case it might be valuable to study combinations and self-adaptive local transition between the models. A simple combination, preserving attractive theoretical properties is given by

$$u_t - \overline{u} \times (\nabla \times \overline{u}) + \nabla q - \nu \Delta u = f, \qquad (7.18)$$

$$\nabla \cdot \overline{u} = 0, \text{ and } -\delta^2 \Delta \overline{u} + \overline{u} = u. \qquad (7.19)$$

An interesting possibility is to include a switching parameter, $0 \leq \theta(x, t) \leq 1$, and consider

$$u_t - [\theta \overline{u} + (1 - \theta)u] \times \nabla \times [\theta u + (1 - \theta)\overline{u}] + \nabla q - \nu \Delta u = f, \qquad (7.20)$$

$$\nabla \cdot [\theta \overline{u} + (1 - \theta)u] = 0, \text{ and } -\delta^2 \Delta \overline{u} + \overline{u} = u. \qquad (7.21)$$

This possibility must include determination of a method for self-adaptively switching between models locally, i.e., determining θ.

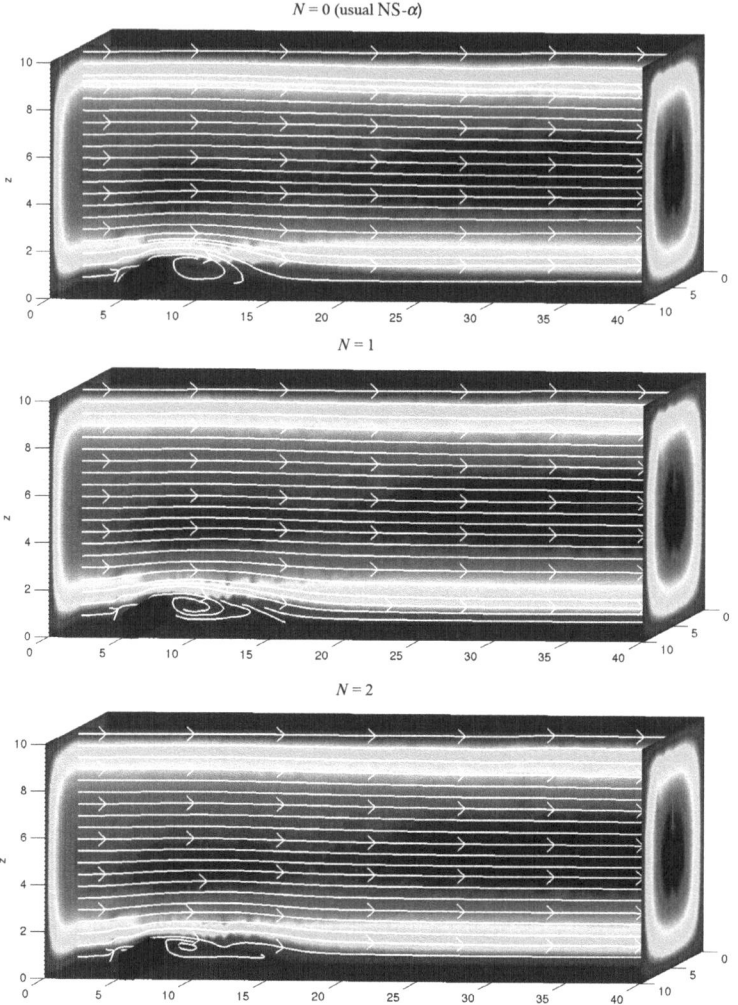

Fig. 7.3 Shown above is velocity streamlines and speed contours on the $x = 5$ sliceplanes of the $T = 10$ velocity solutions for 3D flow over a step, for NS-α-deconvolution, with $N = 0$, 1, 2, from top to bottom

7.6.2 Scale Truncation, Eddy Viscosity, VMMs and Time Relaxation

One basic difficulty with many regularizations is that scale truncation is insufficient. While the model's micro-scales, η_{model}, (as predicted by turbulence phenomenology) is larger that the NSE microscale, η_{NSE}, it is far larger than the filter length scale:

$$\eta_{model} >> \delta \simeq O(h).$$

In the usual K41 phenomenology, [F95], micro-scales can only be altered by increasing energy dissipation. There are three possibilities (the last two have less influence on the resolved scales) which can be added to any model to enhance scale truncation: (a) use algorithms which include extra numerical dissipation (which if not carefully done acts on all scales), (b) subgrid eddy viscosity/Variational Multiscale small-small Smagorinsky, and (c) time relaxation.

Using rotation form in NS-alpha and NS-omega requires the use of the Bernoulli pressure, which is generically significantly more complex than the usual pressure. With the use of $grad - div$ stabilization, the inaccuracy in the Bernoulli pressure associated with using rotation form seems to be localized in the pressure error and have much reduced (or even minimal) effect upon the velocity error.

7.7 Remarks

The NS-alpha model is an idea of Holm, Foias, Titi and their collaborators. Its literature (cited throughout this chapter) is very large, beyond what can be surveyed herein and deserves a book of its own. The presentation of the NS-alpha model as a helicity correction to the Leray regularization is from [Reb07] and the NS-alpha deconvolution model is from [RS10]. The NS-omega model is from [LMNR09] (upon which this chapter is based). For further work on NS-omega see [LLMN08,L10,LST08], and [MNOR11]. Numerical analysis of NS alpha includes [Cag10, Con10] and [MR10]. It has been nontrivial to obtain high accuracy methods for NS alpha (possibly with deconvolution) for non periodic BCs. This was achieved in the last paper. It is insightful to consider a parallel difficulty for the Leray-alpha model and the modified Leray model of Ilyin, Lunasin and Titi, [ILT05]. The Leray and modified Leray regularizations are given (respectively) by $\nabla \cdot u = 0$ and

$$u_t + \begin{pmatrix} \overline{u} \cdot \nabla u \\ \text{or} \\ u \cdot \nabla \overline{u} \end{pmatrix} - \nu \triangle u + \nabla p = f(x,t).$$

The Leray regularization is L^2 stable under no slip BCs. (Take the inner product of the Leray model with u.) Stability is proven for the modified Leray regularization under periodic BCs by taking the inner product with $\left(-\delta^2 \triangle + 1 \right)^{-1} u$. The result gives stability is a weaker norm (equivalent to the norm in H^{-1}). Adapting the stability proof to non-periodic BCs is also nontrivial. Very recent tests reported by Geurts comparing both as a basis for numerical computations (not as theoretical tools) have been consistent with these differences: the Leray model was superior and the modified Leray more sensitive. Let us now import this observation to the NSE with nonlinearity in rotation form. The NS-alpha and omega models are given by, respectively, $\nabla \cdot u = 0$ and

$$u_t + \begin{pmatrix} \omega \times \overline{u} \\ or \\ \overline{\omega} \times u \end{pmatrix} - \nu \triangle u + \nabla P = f(x,t).$$

Stability in a weak H^{-1} like norm is established for NS-alpha by taking the inner product of NS-alpha with $\left(-\delta^2 \triangle + 1\right)^{-1} u$. Stability in L^2 for NS-omega is established by taking the inner product of NS-omega with u. Given this difference and the reported tests of the Leray family, it is not surprising that robust numerical methods for NS-omega are easy to obtain while those for NS-alpha require care.

Beyond the similarities with the Leray family, both NS-alpha and NS-omega use the rotational form of the NSE nonlinearity (in their simplest form). The rotational form also requires special care, even for the NSE. The work presented on this topic was based on [LMNOR09]; see also [O99, O02, OR02, OR04, LO02] for further work on this topic.

"It would appear that we have reached the limits of what is possible to achieve with computer technology, although one should be careful with such statements, as they tend to sound pretty silly in 5 years."
    J. von Neumann, 1949.

Appendix A
Deconvolution Under the No-Slip
Condition and the Loss of Regularity

Most of the estimates for filtering and deconvolution errors have been based on periodic boundary conditions. In this appendix we survey what can be proven for Dirichlet boundary conditions. Let Ω be a bounded, regular domain in $\mathbb{R}^d, d = 1, 2, 3$ with C^{k+2} boundary and $0 < \delta \leq 1$ a small parameter. Consider the elliptic-elliptic singular perturbation problem

$$- \delta^2 \triangle u + u = f, \text{ in } \Omega, \tag{A.1}$$

$$u = 0, \text{ on } \partial\Omega. \tag{A.2}$$

Let $H^k = H^k(\Omega)$ denote the Sobolev space of all functions with derivatives of order $\leq k$ in $L^2(\Omega) = H^0$ with associated norm $|| \cdot ||_k$ and semi-norm $| \cdot |_k$. If $k = 0$ we drop the subscript 0 in the norm and write simply $|| \cdot ||$. The Sobolev space $H_0^1(\Omega)$ is $H_0^1 := \{v \in H^1 : v = 0 \text{ on } \partial\Omega\}$. For (A.1), we assume

$$f \in H^k(\Omega) \bigcap H_0^1(\Omega). \tag{A.3}$$

In particular, we stress that *this implies the important condition*

$$f = 0 \text{ on } \partial\Omega. \tag{A.4}$$

This condition precludes simple boundary layers in u but does not imply higher derivatives of u are free of layers. From (A.1), it also implies that $\triangle u = 0$ on $\partial\Omega$.

The shift theorem, e.g., [GT], implies that the solution of (A.1)-(A.2) satisfies

$$u \in H^{k+2}(\Omega) \bigcap H_0^1(\Omega), \text{ and } ||u||_{k+2} \leq C(\delta)||f||_k,$$

where $C(\delta) \to \infty$ as $\delta \to 0$. Herein we investigate the question of *uniform in* δ regularity theorem: whether there is a $C = C(k, \Omega)$, independent of δ, such that the solution of (A.1)-(A.2) under (A.3) satisfies

W.J. Layton and L.G. Rebholz, *Approximate Deconvolution Models of Turbulence*, Lecture Notes in Mathematics 2042, DOI 10.1007/978-3-642-24409-4, © Springer-Verlag Berlin Heidelberg 2012

$$||u||_k \leq C||f||_k. \tag{A.5}$$

This simple question has turned out to be more delicate than it appeared at first. We prove the following in Sect. A.1.

Theorem 66. *Suppose* $f \in H^2(\Omega) \bigcap H_0^1(\Omega)$. *Then*

$$||u||_l \leq C||f||_l, \ for \ l = 0, 1, 2. \tag{A.6}$$

If $f \in H^4(\Omega) \bigcap H_0^1(\Omega), \triangle f \in H_0^1(\Omega)$. *Then*

$$||u||_l \leq C||f||_l, \ for \ l = 0, 1, 2, 3, 4. \tag{A.7}$$

In general, suppose $f \in H^{2k}(\Omega) \bigcap H_0^1(\Omega), \triangle^j f \in H_0^1(\Omega), j = 1, \cdots, k - 1$. *Then for* $l = 1, \cdots, 2k$

$$||u||_l \leq C||f||_l. \tag{A.8}$$

The assumption that powers of $\triangle f$ vanish on the boundary can be weakened in appearance to read that normal derivatives of the same order vanish on the boundary, Sect. A.2. However, examples in Sect. A.3 show that an extra condition is necessary. Without it we have the following (Sect. A.2).

Corollary 67. *Suppose* $f \in H^3(\Omega) \bigcap H_0^1(\Omega)$. *Then*

$$||u||_3 \leq C(||f||_3 + \delta^{-1}||f||_2)$$

It seems likely that all the results herein are in the published literature, explicitly or implicitly, somewhere (e.g., Lions [L73]). Simple examples in $1d$ show that Theorem 66 cannot be true if (A.4) does not hold and that local versions of Theorem 66 cannot hold as well.

A.1   Regularity by Direct Estimation of Derivatives

We consider first the cases where results may be proven by a direct argument. The estimates in Lemma 68 likely appear in every paper on regularity of (A.1)-(A.2). The estimates in Lemma 69 appear in Tartar [T93]. All constants are uniform in δ.

Lemma 68. *Under* (A.3) *we have*

$$\delta^2||\triangle u|| + \delta||\nabla u|| + ||u|| \leq C||f||, \tag{A.9}$$

$$\delta||\triangle u|| + ||\nabla u|| \leq C||\nabla f||. \tag{A.10}$$

Proof. Multiply the equation (A.1) by u and integrate over the domain Ω. This gives

$$\delta^2||\nabla u||^2 + ||u||^2 = (f, u) \le \frac{1}{2}||f||^2 + \frac{1}{2}||u||^2,$$

and the first estimate follows. For the second estimate, the equation (A.1) implies $-\delta^2 \triangle u = f - u$ and thus

$$\delta^2||\triangle u|| \le ||f|| + ||u|| \le C||f||.$$

The gradient bound in (A.9) is improvable. Indeed, if $f \in H_0^1(\Omega)$ then

$$u \in H^3(\Omega) \bigcap H_0^1(\Omega).$$

Multiply by $-\triangle u$ and integrate. This gives

$$\delta^2||\triangle u||^2 + ||\nabla u||^2 = (\nabla f, \nabla u) \le \frac{1}{2}||\nabla f||^2 + \frac{1}{2}||\nabla u||^2,$$

and the second estimate (A.10) follows. $\qquad\square$

Next we use a reformulation of (A.1) in Tartar [T93]. This reformulation is critical for the next step and the regularity question can be restated for the reformulation. Since $f \in H_0^1(\Omega)$,

$$(u - f) \in H_0^1(\Omega).$$

Subtraction gives the equation

$$- \delta^2 \triangle(u - f) + (u - f) = \delta^2 \triangle f, \text{ in } \Omega, \qquad (A.11)$$

$$(u - f) = 0, \text{ on } \partial\Omega. \qquad (A.12)$$

Lemma 69. *There is a C independent of δ such that*

$$||u||_k \le C||f||_k$$

if and only if

$$||u - f||_k \le C||f||_k.$$

Proof. This follows from the triangle inequality:

$$||u||_k \le ||u - f||_k + ||f||_k,$$

$$||u - f||_k \le ||f||_k + ||u||_k.$$

$\qquad\square$

Consider therefore problem (A.11)-(A.12).

Lemma 70. *Under (1.2) we have*

$$||\nabla(u - f)|| \leq C||\nabla f||$$
$$||u - f|| \leq C\delta||\nabla f||,$$
$$||\nabla(u - f)|| \leq C\delta||\triangle f||,$$
$$||u - f|| \leq C\delta^2||\triangle f||,$$
$$||\triangle(u - f)|| \leq C||\triangle f||.$$

Proof. Multiply (A.11) by $u - f$ and integrate. This gives, after the usual manipulations,

$$\delta^2||\nabla(u - f)||^2 + ||u - f||^2 = (\delta^2\triangle f, u) \leq \delta^2||\nabla(u - f)||||\nabla f||.$$

This proves the first two estimates. On the above RHS we can also write $(\delta^2\triangle f, u) \leq \frac{\delta^4}{2}||\triangle f||^2 + \frac{1}{2}||u - f||^2$, which proves the third and fourth estimates. Equation (A.11) now implies

$$\delta^2||\triangle(u - f)|| \leq \delta^2||\triangle f|| + ||u - f|| \leq C\delta^2||\triangle f||.$$

Thus,

$$||\triangle(u - f)|| \leq C||\triangle f||.$$

$\square$

Since $\partial\Omega$ is smooth, $||\triangle u||$ and $|u|_2$ are equivalent. Thus we have the following.

Corollary 71. *For $k = 0, 1, 2$ we have*

$$||u||_k \leq C||f||_k.$$

These simple estimates can be continued and give precise information. If

$$f \in H_0^1(\Omega), \text{ then } u \in H^3(\Omega) \bigcap H_0^1(\Omega)$$

and from the equation $-\delta^2\triangle u = f - u \in H_0^1(\Omega)$, so

$$\triangle u \in H_0^1(\Omega).$$

Thus $\triangle u$ satisfies

$$-\delta^2\triangle\triangle u + \triangle u = \triangle f \ \in H_0^1(\Omega), \text{ in } \Omega, \tag{A.13}$$
$$\triangle u = 0, \text{ on } \partial\Omega.$$

Applying the above estimates to $\triangle u$ gives for $k = 0, 1, 2, 3, 4$

$$||u||_k \leq C||f||_k. \tag{A.14}$$

Clearly this can be repeated. Repeating this argument proves Theorem 66.

Theorem 72. *Suppose* $f \in H^{2k}(\Omega) \bigcap H_0^1(\Omega), \triangle^j f \in H_0^1(\Omega), j = 1, \ldots,$ $k - 1$. *Then for* $l = 1, \cdots, 2k$

$$||u||_l \leq C||f||_l. \tag{A.15}$$

A.2   The Bootstrap Argument

The usual path to regularity is via a bootstrap argument. The section considers how the classical bootstrap argument applies to the regularity issue. We give a different proof of the basic regularity result of the last section. We return to the case:

$$f \in H^k(\Omega) \bigcap H_0^1(\Omega). \tag{A.16}$$

A.2.1   The Case $k = 3$

The usual procedure, [GT] is first to use a partition of unity. Then, change variables to locally flatten the boundary. The sought estimate is first proven for tangential derivatives through tangential difference quotients. Finally, the last derivative in the normal direction is bounded by tangential derivatives through the equation. This section uses an alternate but related strategy (which was suggested to the author by L. Tartar and used in [T93]) that simplifies the argument considerably. The key is the following observation.

A.2.2   Observation

Let L *be a smooth, first order differential operator which acts tangentially to* $\partial\Omega$. *Then, for any*

$$v \in H^2(\Omega) \bigcap H_0^1(\Omega)$$

we have

$$Lv \in H_0^1(\Omega), \text{ and}$$

$$\triangle Lv = L\triangle v + Av,$$

where A *is a second order differential operator.*

Proposition 73. *Suppose L is a smooth, first order differential operator which acts tangentially to $\partial\Omega$. Then,*

$$||\triangle L(u - f)|| \le C||f||_3$$

Proof. Applying L to the equation

$$-\delta^2\triangle(u - f) + (u - f) = \delta^2\triangle f, \text{ in } \Omega,$$
$$(u - f) = 0, \text{ on } \partial\Omega.$$

gives

$$- \delta^2\triangle L(u - f) + L(u - f) = \delta^2 L\triangle f + \delta^2 A(u - f), \text{ in } \Omega, \quad (A.17)$$
$$L(u - f) = 0, \text{ on } \partial\Omega.$$

Apply Lemma (69) to (A.17). This gives

$$||L(u - f)|| \le C\delta^2\{||L\triangle f|| + ||A(u - f)||\},$$

and since A is second order, Lemma (68) implies $||A(u-f)|| \le C||f||_2$. Thus,

$$||L(u - f)|| \le C\delta^2||f||_3.$$

Next, apply the idea in the proof of Lemma (68) to equation (A.17). This gives

$$\delta^2||\triangle L(u - f)|| \le \delta^2||f||_3 + ||L(u - f)|| \le C\delta^2||f||_3.$$

Thus we have

$$||\triangle L(u - f)|| \le C||f||_3$$

for any first order differential operator acting tangentially to the boundary.
$\square$

There remains only to check the norm of the one third order differential operator acting normal to $\partial\Omega$. This is one additive term in $\nabla\triangle(u - f)$ so that the theorem will hold if $||\triangle(u - f)||_1 \le C||f||_3$. To verify this estimate, recall that the equation

$$-\delta^2\triangle(u - f) + (u - f) = \delta^2\triangle f, \text{ in } \Omega,$$
$$(u - f) = 0, \text{ on } \partial\Omega.$$

implies

$$\delta^2||\triangle(u - f)||_1 \le \delta^2||\triangle f||_1 + C||\nabla(u - f)||. \quad (A.18)$$

At this point, the best estimate of the last term in Lemma (69) is

$$\|\nabla(u - f)\| \leq C\delta\|\triangle f\|.$$

This gives the following.

Corollary 74. *Suppose* $f \in H^3(\Omega) \bigcap H_0^1(\Omega)$. *Then*

$$\|u\|_3 \leq C(\|f\|_3 + \delta^{-1}\|f\|_2)$$

To eliminate the δ^{-1} it suffices that $\triangle f = 0$ on $\partial\Omega$.

Lemma 75. *Suppose* $\triangle f \in H_0^1(\Omega)$, *then*

$$\|\nabla(u - f)\| \leq C\delta^2\|\triangle f\|_1. \tag{A.19}$$

Proof. Begin with the equation

$$-\delta^2\triangle(u - f) + (u - f) = \delta^2\triangle f, \text{ in } \Omega,$$
$$(u - f) = 0, \text{ on } \partial\Omega.$$

Taking the inner product with $-\triangle(u - f)$ gives

$$\delta^2\|\triangle(u-f)\|^2 + \|\nabla(u-f)\| = \delta^2(\triangle f, -\triangle(u-f)) \leq \delta^2\|\triangle f\|_1\|\triangle(u-f)\|_{-1}.$$

The key step depends upon the extra regularity $\triangle f \in H_0^1(\Omega)$. With this we can use the estimate

$$(\triangle f, -\triangle(u - f)) \leq \|\triangle f\|_1\|\triangle(u - f)\|_{-1}.$$

Now $(u - f) \in H_0^1(\Omega)$ so $\triangle(u - f) \in H^{-1}(\Omega)$ and

$$\|\triangle(u - f)\|_{-1} \leq C\|\nabla(u - f)\|.$$

Thus,

$$\|\nabla(u - f)\|^2 \leq C\delta^2\|\triangle f\|_1\|\nabla(u - f)\|,$$

completing the proof.   □

It is clear that all that is really needed is that the second normal derivative of the RHS be well defined and vanish on the boundary.

Corollary 76. *Suppose* $f \in H^k(\Omega) \bigcap H_0^1(\Omega)$, *and* $\triangle f \in H_0^1(\Omega)$. *Then, for* $k = 0, 1, 2, 3$ *we have*

$$\|u\|_k \leq C\|f\|_k.$$

A.3   Examples

Example 77 (When $f \neq 0$ on the boundary). This is an example in which $f \in H^1(\Omega)$ from [L73], page 133. First note that the estimates derived imply that $u \to f$ as $\delta \to 0$ weakly in $L^2(\Omega)$. If the RHS does not vanish on the boundary the gradients cannot converge strongly since $H_0^1(\Omega)$ is a closed subspace of $H^1(\Omega)$.

The following example illustrates this. Consider the $1d$ problem

$$-\delta^2 u'' + u = e^{-x}, \text{ in } (0, \infty),$$

$$u = 0, \text{ at } x = 0.$$

The solution is

$$u(x) = \frac{1}{1 - \delta^2} e^{-x} - \frac{1}{1 - \delta^2} e^{-\frac{x}{\delta}}.$$

It is easy to verify that on subdomains away from $x = 0$ there is no difficulty: $u(x) \to f(x)$. The derivatives do not converge near $x = 0$ due to the layer at $x = 0$.

Example 78 (Regularity is false in general). The second example is due to P. Rabier and shows that (A.5) cannot hold for all k without extra conditions on the RHS. Consider (A.1) in one dimension, which reduces to

$$-\delta^2 u'' + u = f, \text{ in } (0, 1),$$

$$u = 0, \text{ at } x = 0, 1.$$

Pick $f \in C^\infty(0, 1)$ with $f''(x) \neq 0$ at all x. In the equation let $x \to 0$. This implies

$$|u''(0)| = \delta^{-2}|f(0)| = 0.$$

Differentiate twice and repeat this argument. We have $-\delta^2 u'''' + u'' = f''$, in $(0, 1)$. Thus, at $x = 0$

$$|u''''(0)| = \delta^{-2}|f''(0)| \to \infty, \text{ as } \delta \to 0.$$

By the Sobolev theorem we have

$$|u''''(0)| \leq C\|u\|_5 \nleq C\|f\|_k$$

for any k (in particular $k = 5$) since the LHS blows up while the RHS is bounded.

Example 79. The following $1d$ example, due to Xinfu Chen, connects the regularity question to results in asymptotic analysis. Suppose $\Omega = (0, 1)$ and $I = [a, b]$ is properly contained in $(0, 1)$ so that $0 < a < b < 1$. Let $f(x) \equiv 1$ on I and $\to 0$ smoothly off I. Consider (1.1) in one dimension, which reduces to

$$-\delta^2 u'' + u = f, \text{ in } (0,1), \tag{A.20}$$

$$u = 0, \text{ at } x = 0, 1. \tag{A.21}$$

Differentiating (A.21) and setting $x \in I$, we have

$$u''' = \frac{1}{\delta^2} u',$$

$$u'''' = \delta^{-2} u'' = \delta^{-4} u'$$

and thus

$$u^{(2k)} = \delta^{-2k} u' \text{ on } I.$$

Now, if the Theorem 1.1 holds we must have the apparently impossible relation

$$\delta^{-2k} ||u'||_{L^2(I)} = ||u^{(2k)}||_{L^2(I)} \le ||u^{(2k)}||_{L^2(0,1)} \le C(k) ||f||_{2k}. \tag{A.22}$$

From (4.2) it appears that uniform in δ regularity is impossible. However, the solution to (1.1) (and thus (4.1)) is an approximation to $f(x)$ and thus since this particular function satisfies $f'(x) = 0$ on I we should have $u'(x)$ small there as well. Asymptotic analysis of (3.1) indicates that on I, $u'(x)$ is, in fact, exponentially close to $f'(x)$ and thus is exponentially close to zero, e.g., [B75, B79, E79]. Thus, $\sup_{0 < \delta \le 1} \delta^{-2k} ||u'||_{L^2(I)} \le C(k)$, which is consistent with the possibility of uniform regularity. Indeed, if $f(x)$ is extended to have compact support then the regularity result does hold while it fails for other smooth extensions by Example 2.

Indeed, this last example cannot be a counterexample because the same local argument would apply to the same problem under periodic boundary conditions (where the RHS is extended periodically). In the periodic case we can verify uniform regularity by direct calculation. Indeed, we calculate

$$u(x) = \sum_{j \in \mathbb{Z}} \frac{1}{1 + \delta^2 (\frac{j}{2\pi})^2} f_j e^{ijx/2\pi},$$

where f_j are the Fourier coefficients of $f(x)$. We calculate further that in the periodic case uniform regularity holds with $C(k) = 1$ trivially since:

$$||u^{(k)}||^2 = \sum_{j \in \mathbb{Z}} \left[\frac{1}{1 + \delta^2 (\frac{j}{2\pi})^2} \right]^2 (\frac{j}{2\pi})^{2k} |f_j|^2 \le$$

$$\le \sum_{j \in \mathbb{Z}} (\frac{j}{2\pi})^{2k} |f_j|^2 = ||f^{(k)}||^2.$$

A.4   Application to Differential Filters

As noted above, this was motivated by an application to Germano's idea of using differential filters as a basic for large eddy simulation. Given (typically a fluid velocity) $\phi \in H^k(\Omega) \cap H_0^1(\Omega)$, its differential filter $\bar{\phi}$ is the unique solution of

$$A\bar{\phi} := -\delta^2 \triangle \bar{\phi} + \bar{\phi} = \phi, \text{ in } \Omega, \qquad (A.23)$$
$$\bar{\phi} = 0, \text{ on } \partial\Omega.$$

The following question occurs in the analysis of the accuracy of approximate deconvolution operators:

For what n and k do we have

$$||A^{-n}\phi||_k \leq C||\phi||_k$$

uniformly in δ?

We trace through now some answers provided by Theorem 66. For $n = 1$ Theorem 66 implies

$$||\bar{\phi}||_k \leq C||\phi||_k, \text{ for } k = 0, 1, 2 \text{ and that}$$
$$\triangle \bar{\phi} = 0 \text{ on } \partial\Omega.$$

Since $\triangle \bar{\phi} = 0$ on $\partial\Omega$, we can apply a higher estimate to the case $n = 2$. Indeed, $A^{-2}\phi = A^{-1}\bar{\phi} = \bar{\bar{\phi}}$ so that

$$||\bar{\bar{\phi}}||_k \leq C||\bar{\phi}||_k, \text{ for } k = 0, 1, 2, 3, 4 \text{ and that}$$
$$\triangle \bar{\bar{\phi}} = \triangle \bar{\phi} = 0 \text{ on } \partial\Omega.$$

Further, the equation for $\bar{\bar{\phi}}$ is

$$-\delta^2 \triangle \bar{\bar{\phi}} + \bar{\bar{\phi}} = \bar{\phi}, \text{ in } \Omega. \qquad (A.24)$$

Taking the Laplacian of this equation gives

$$-\delta^2 \triangle^2 \bar{\bar{\phi}} + \triangle \bar{\bar{\phi}} = \triangle \bar{\phi}, \text{ in } \Omega. \qquad (A.25)$$

Now, let $x \to \partial\Omega$ and use $\triangle \bar{\bar{\phi}} = \triangle \bar{\phi} = 0$ on $\partial\Omega$. This implies

$$\triangle^2 \bar{\bar{\phi}} = \triangle \bar{\bar{\phi}} = \bar{\bar{\phi}} = 0 \text{ on } \partial\Omega$$

so that even higher uniform regularity can be inferred for $\overline{\overline{\overline{\phi}}}$.

$$||\overline{\overline{\overline{\phi}}}||_k \leq C||\overline{\overline{\phi}}||_k, \text{ for } k = 0, 1, 2, 3, 4, 5, 6.$$

This argument can be continued.

A.5 Remarks

This section is based on [L07] which also treats the nonlinear case. Patrick Rabier, Catalin Trenchea and Luc Tartar gave important help on the estimates in this chapter. The proof of Theorem 66 (for $k > 2$) is due to Patrick Rabier as well as the critical second example of Sect. A.3. The proof of Proposition 73 is due to Luc Tartar and Sect. A.2 is based on a helpful communication of his. Lemmas 68 and 69 are from his paper [T93]. The third example is due to Xinfu Chen and came from a stimulating discussion with him.

References

[AS01]   N. A. ADAMS AND S. STOLZ, *Deconvolution methods for subgrid-scale approximation in large eddy simulation*, Modern Simulation Strategies for Turbulent Flow, R.T. Edwards, 2001.

[AS02]   N. A. ADAMS AND S. STOLZ, *A subgrid-scale deconvolution approach for shock capturing*, Journal of Computational Physics, 178 (2002), 391–426.

[ABHM06] E. AKERVIK, L. BRANDT, D. S. HENNINGSON, J. HOEPFFNER, O. MARXEN, P. SCHLATTER, *Steady solutions of the Navier–Stokes equations by selective frequency damping*, Physics of Fluids, 18, 068102 (2006), 1–4.

[ALP04]   M. ANITESCU, W. LAYTON AND F. PAHLEVANI, *Implicit for local effects and explicit for nonlocal effects is unconditionally stable*, ETNA, 18 (2004), 174–187.

[Arc02]   ARCHIMEDES, T. L. HEATH (Translator), *The Works of Archimedes*, Dover, (2002).

[B76]   G. BAKER, *Galerkin Approximation for the Navier–Stokes Equations*, Report, Harvard University, (1976).

[B79]   J. BARANGER, *On the thickness of the boundary layer in elliptic-elliptic singular perturbation problems*, 395–400 in: *Numerical Analysis of Singular Perturbation Problems* (P.W. Hemker and J.J.H. Miller, eds.) Academic press, NY, 1979.

[Bar83]   J. BARDINA, *Improved turbulence models based on large eddy simulation of homogeneous, incompressible turbulent flows*, Ph.D. thesis, Stanford University, Stanford, (1983).

[BFG02]   S. BASU, E. FOUFOULA-GEORGIOU AND F. PORTE-AGEL, *Predictability of atmospheric boundary layer flows as a function of scale*, UMn Report, UMSI 2002/89, (2002).

[BGJ07]   L.C. BERSELLI, C.R. GRISANTI, AND V. JOHN, *Analysis of commutation errors for functions with low regularity*, J. Comput. Appl. Math. 206 (2007), 1027–1045.

[BIL06]   L. C. BERSELLI, T. ILIESCU, AND W. LAYTON, *Mathematics of Large Eddy Simulation of Turbulent Flows*. Springer, Berlin, (2006).

[BJG07]   L.C. BERSELLI, V. JOHN AND C. GRISANTI, *Analysis of commutation errors for functions with low regularity*, J. Comput. Appl. Math., 206 (2007), 1027–1045.

[BL11]   L.C. BERSELLI AND R. LEWANDOWSKI, *Convergence of approximate deconvolution models to the filtered Navier–Stokes Equations*, under revision in Ann. IHP, 2011

[BB98]   M. BERTERO AND B. BOCCACCI, *Introduction to Inverse Problems in Imaging*, IOP Publishing Ltd., (1998).

W.J. Layton and L.G. Rebholz, *Approximate Deconvolution Models of Turbulence*, Lecture Notes in Mathematics 2042, DOI 10.1007/978-3-642-24409-4, © Springer-Verlag Berlin Heidelberg 2012

[B75]   J.G. BESJES, *Singular perturbation problems for linear elliptic differential operators of arbitrary order, I. Degeneration to elliptic operators*, JMAA 49(19795) 24–46.

[BBJL07] M. BRAACK, E. BURMAN, V. JOHN, AND G. LUBE, *Stabilized finite element methods for the generalized Oseen problem*, Comput. Methods Appl. Mech. Eng., 196, (2007), 853–866.

[BS94] S. BRENNER AND L.R. SCOTT, *The Mathematical Theory of Finite Element Methods*, Springer-Verlag, 1994.

[BR10] A. BOWERS AND L. REBHOLZ, *Increasing accuracy and efficiency in FE computations of the Leray-deconvolution model*, Numerical Methods for Partial Differential Equations, to appear, (2010).

[BLCR10] A. BOWERS, B. COUSINS, A. LINKE AND L. REBHOLZ, *New connections between finite element formulations of the Navier–Stokes equations*, Journal of Computational Physics, 229 (2010), 9020–9025.

[Cag10] A. CAGLAR, *Convergence analysis of the Navier–Stokes-alpha model*, Numerical methods for partial differential equations, 26 (2010), 1154–1167.

[CHQZ88] C. CANUTO, M. HUSSAINI, A. QUARTERONI, AND T. ZANG, *Spectral methods in fluid dynamics*, Springer-Verlag Inc., New York, 1988.

[CCE03] Q. CHEN, S. CHEN, AND G. EYINK, *The joint cascade of energy and helicity in three dimensional turbulence*, Physics of Fluids, 15(2) (2003), 361–374.

[CHOT05] A. CHESKIDOV, D. D. HOLM, E. OLSON AND E. S. TITI, *On a Leray-α model of turbulence*, Royal Society London, Proceedings, Series A, Mathematical, Physical and Engineering Sciences, 461 (2005), 629–649.

[C98]   P. COLETTI, *Analytical and numerical results for k-epsilon and large eddy simulation turbulence models*, Ph.D. Thesis, UTM-PHDTS 17, U. Trento, 1998.

[Con10] J. CONNORS, *Convergence analysis and computational testing of a finite element discretization of the Navier–Stokes alpha model*, Numerical Methods for Partial Differential Equations, 26(6) (2010), 1328–1350.

[CHL09] J. M. CONNORS, J. S. HOWELL AND W. LAYTON, *Decoupled time stepping methods for fluid-fluid interaction*, submitted to SINUM, 2009.

[CL10] J. CONNORS AND W. LAYTON, *On the accuracy of the finite element method plus time relaxation*, Math. Comp. 79 (2010), 619–648.

[CRW10] B. COUSINS, L. REBHOLZ AND N. WILSON, *Enforcing energy, helicity and strong mass conservation in FE computations for incompressible Navier–Stokes simulations*, submitted, (2011).

[DB86] Y. M. DAKHOUL AND K. W. BEDFORD, *Improved averaging method for turbulent flow simulation. Part I: Theoretical development and application to Burger's transport equation*, Int. J. Numer. Methods Fluids, 6 (1986), p. 49.

[DM01] A. DAS AND R.D. MOSER, *Filtering boundary conditions for LES and embedded boundary simulations*, DNS/LES progress and challenges (C. Liu, L. Sakeland, and T. Beutner, eds.), Greyden Press, Columbus, 2001, pp. 389–396.

[Day90] M. A. DAY, *The no-slip condition of fluid dynamics*, Erkenntnis, 33 (1990), 285–286.

[DG01a] P. DITLEVSEN AND P. GIULIANI, *Cascades in helical turbulence*, Physical Review E, 63 (2001), 1–4.

[DG01b] P. DITLEVSEN AND P. GIULIANI, *Dissipation in helical turbulence*, Physics of Fluids, 13(11) (2001), 3508–3509.

[DG95] C. DOERING AND J.D. GIBBON, *Applied analysis of the Navier–Stokes equations*, Cambridge, (1995).

[DG91] Q. DU AND M. GUNZBURGER, *Analysis of a Ladyzhenskaya model for incompressible viscous flow*, JMAA 155 (1991), 21–45.

[D04]   A. DUNCA, *Space averaged Navier–Stokes equations in the presence of walls*, Ph.D. Thesis, University of Pittsburgh, 2004.

[DE06] A. DUNCA AND Y. EPSHTEYN, *On the Stolz-Adams deconvolution model for the Large-Eddy simulation of turbulent flows*, SIAM J. Math. Anal., 37(6) (2006), 1890–1902.

[D03] A. DUNCA, *Optimal design of fluid flow using subproblems reduced by large eddy simulation*, Technical Report ANL/MCS (2003).

[DJL04] A. DUNCA, V. JOHN AND W. LAYTON, *The commutation error of the space averaged Navier–Stokes equations in a bounded domain*, in: G.P. Galdi, J.G. Heywood, R. Rannacher (Eds.), Contributions to Current Challenges in Mathematical Fluid Mechanics, Advances in Mathematical Fluid Mechanics 3, Birkhauser Verlag Basel, (2004), 53–78.

[E79] W. ECKHAUS, *Asymptotic analysis of singular perturbations*, N. Holland, Amsterdam,1979.

[ELN06] V. ERVIN, W. LAYTON AND M. NEDA, *Numerical analysis of a higher order time relaxation model of fluids*, Int. J. Numer. Anal. and Modeling, 4(3–4) (2007), 648–670.

[Fe00] C. L. FEFFERMAN, *Official Clay prize problem description: Existence and smoothness of the Navier–Stokes equations*, http://www.claymath.org/millennium/, (2000).

[F97] C. FOIAS, *What do the Navier–Stokes equations tell us about turbulence?* Contemporary Mathematics, 208 (1997), 151–180.

[FHT01] C. FOIAS, D. D. HOLM AND E. S. TITI, *The Navier–Stokes-alpha model of fluid turbulence*, Physica D, (152–153) (2001), 505–519.

[FHT02] C. FOIAS, D. HOLM AND E. TITI, *The three dimensional viscous Camassa-Holm equations, and their relation to the Navier–Stokes equations and turbulence theory*, Journal of Dynamics and Differential Equations, 14 (2002), 1–35.

[F95] U. FRISCH, *Turbulence*, Cambridge, (1995).

[Ga00] G. P. GALDI, *Lectures in Mathematical Fluid Dynamics*, Birkhauser-Verlag, (2000).

[Gal94] G.P.GALDI, *An introduction to the Mathematical Theory of the Navier–Stokes equations, Volume I*, Springer, Berlin, (1994).

[GL00] G. P. GALDI AND W. J. LAYTON, *Approximation of the large eddies in fluid motion II: A model for space-filtered flow*, Math. Models and Methods in the Appl. Sciences, 10 (2000), 343–350.

[Ger86] M. GERMANO, *Differential filters of elliptic type*, Phys. Fluids, 29 (1986), 1757–1758.

[GPMC91] M. GERMANO, U. PIOMELLI, P. MOIN, W. CABOT, *A dynamic subgrid-scale eddy viscosity model*, Physics of Fluids A3 (1991) 1760–1765.

[Geu97] B. J. GEURTS, *Inverse modeling for large eddy simulation*, Phys. Fluids, 9 (1997), 3585.

[G03] B. J. GEURTS, *Elements of direct and large eddy simulation*, Edwards Publishing, (2003).

[GH05] B. J. GEURTS AND D. D. HOLM, *Leray and LANS-alpha modeling of turbulent mixing*, J. of Turbulence, 00(2005), 1–42.

[GH03] B. J. GEURTS AND D. D. HOLM, *Regularization modeling for large eddy simulation*, Physics of Fluids, 15(1) (2003), 13–16.

[GT] D. GILBARG AND N.S. TRUDINGER, *Elliptic partial differential equations of second order*, Springer, Berlin, (2001).

[GS98] P. GRESHO AND R. SANI, *Incompressible flow and the finite element method*, Wiley, (1998).

[Gue04] R. GUENANFF, *Non-stationary coupling of Navier–Stokes/Euler for the generation and radiation of aerodynamic noises*, Ph.D. thesis, Dept. of Mathematics, Universite Rennes 1, Rennes, France, (2004).

[Guer] J.-L. GUERMOND, *Subgrid stabilization of Galerkin approximations of monotone operators*, C. R. Acad. Sci. Paris, S érie I, 328(7) (1999), 617–622.

[GOP03] J.L. GUERMOND, S. PRUDHOMME AND J.T. ODEN, *An interpretation of the NS alpha model as a frame indifferent Leray regularization*, Physica D Nonlinear Phenomena, 177 (2003), 23–30.

[GP05] J.-L. GUERMOND AND S. PRUDHOMME, *On the construction of suitable solutions of the Navier–Stokes equations and questions regarding the definition of large eddy simulation*, Physica D, 207 (2005) 64–78.

[G89] M.D. GUNZBURGER, *Finite Element Methods for Viscous Incompressible Flows - A Guide to Theory, Practices, and Algorithms*, Academic Press, (1989).

[HV03] HASELBACHER, A. AND VASILYEV, O.V., *Commutative discrete filtering on unstructured grids based on least-squares techniques*, Journal of Computational Physics, 187(1) (2003), 197–211.

[Horiuti87] K. HORIUTI, *Comparison of conservative and rotation forms in large eddy simulation of turbulent channel flow*, J.C.P., 71 (1987), 343–370.

[Horiuti98] K. Horiuti AND T. ITAMI, *Truncation error analysis of the rotation form of convective terms in the Navier–Stokes equations*, J.C.P., 145 (1998), 671–692.

[HKJ00] T. HUGHES, L. MAZZEI AND K. JANSEN, *Large Eddy Simulation and the Variational Multiscale Method*, Computing and Visualization in Science, 3 (1/2)(2000) 47–59.

[HOM01] T. HUGHES, A. OBERAI AND L. MAZZEI, *Large Eddy Simulation of Turbulent Channel Flows by the Variational Multiscale Method*, Physics of Fluids, 13(6)(2001) 1784–1799.

[IL98] T. ILIESCU AND W. LAYTON, *Approximating the larger eddies in fluid motion III: The Boussinesq model for turbulent fluctuations,* Analele Stinfice ale Universitatii "Al. I. Cuza" Iasi, XLIV (1998), 245–261.

[ILT05] A. A. ILYIN, E. M. LUNASIN AND E. S. TITI, *A modified Leray-alpha subgrid-scale model of turbulence*, Nonlinearity, 19 (2006), 879–897.

[J04] V. JOHN, *Large Eddy Simulation of Turbulent Incompressible Flows,* Springer, Berlin, (2004).

[JL00] V. JOHN AND W. LAYTON, *Analysis of numerical errors in large eddy simulation*, SINUM, 40 (2000) 995–1020.

[JLS04] V. JOHN, W. LAYTON AND N. SAHIN, *Derivation and analysis of near wall models for channel and recirculating flows*, Computers and Mathematics with Applications, 48 (2004), 1135–1151.

[JL06] V. JOHN AND A. LIAKOS, *Time dependent flow across a step: the slip with friction boundary condition*, Int. J. Numer. Meth. Fluids, 50 (2006), 713–731.

[Koe84] J.J. KOENDERINK, *The structure of images*, Biol. Cybernetics, 50 (1984), 363–370.

[LLMN08] A. LABOVSCHII, W. LAYTON, C. MANICA, M. NEDA, L. REBHOLZ, I. STANCULESCU AND C. TRENCHEA, *Architecture of Approximate Deconvolution Models of Turbulence*, Ercoftac Series: Quality and Reliability of Large-Eddy Simulation, 10.1007/978-1-4020-8578-9_1, Editors: Johan Meyers, Bernard J. Geurts and Pierre Sagaut, (2008).

[L02] W LAYTON, *A connection between subgrid-scale eddy viscosity and mixed methods*, Applied Math and Computing, 133 (2002), 147–157.

[L99] W. LAYTON, *Weak imposition of "no-slip" conditions in finite element methods*, Computers & Mathematics with Applications, 38 (1999), 129–142.

[L08] W. LAYTON, *Introduction to the Numerical Analysis of Incompressible, Viscous Flows*, SIAM, (2008).

[L07] W. LAYTON, *A remark on regularity of elliptic-elliptic singular perturbation problem*, Technical Report, available at http://www.math.pitt.edu/ techreports.html, (2007).

[L10] W LAYTON, *Existence of smooth attractors for the Navier-Stokes-omega model of turbulence*, JMAA, 366, (2010), 81–89.

[L07b] W. LAYTON, *Superconvergence of finite element discretization of time relaxation models of advection*, BIT Numerical Mathematics 47 (2007), 565–576.

[LL02] W LAYTON AND R LEWANDOWSKI, *Analysis of an Eddy Viscosity Model for Large Eddy Simulation of Turbulent Flows*, Journal of Mathematical Fluid Mechanics, 4, 2002, 374–399.

[LL08] W. LAYTON AND R. LEWANDOWSKI, *On the Leray deconvolution model*, Analysis and Applications, 6(1) (2008), 23–49.

[LL03] W. LAYTON AND R. LEWANDOWSKI, *A simple and stable scale similarity model for large eddy simulation: energy balance and existence of weak solutions*, Applied Math. Letters 16 (2003), 1205–1209.

[LL05] W. LAYTON AND R. LEWANDOWSKI, *Residual stress of approximate deconvolution large eddy simulation models of turbulence. Journal of Turbulence*, 46(2) (2006), 1–21.

[LL06a] W. LAYTON AND R LEWANDOWSKI, *On a well posed turbulence model*, Discrete and Continuous Dynamical Systems - Series B, 6 (2006), 111–128.

[LMNR08b] W. LAYTON, C. MANICA, M. NEDA AND L. REBHOLZ, *Numerical analysis of a high accuracy Leray-deconvolution model of turbulence*, N.M.P.D.E, 24(2) (2008), 555–582.

[LMNR08] W. LAYTON, C. MANICA, M. NEDA AND L. REBHOLZ, *The joint Helicity-Energy cascade for homogeneous, isotropic turbulence generated by approximate deconvolution models,* Advances and Applications in Fluid Mechanics, 4(1) (2008), 1–46.

[LMNOR09] W. LAYTON, C. MANICA, M. NEDA, M. OLSHANSKII AND L. REBHOLZ, *On the accuracy of the rotation form in simulations of the Navier–Stokes equations*, Journal of Computational Physics, 228(9) (2009), 3433–3447.

[LMNR09] W. LAYTON, C. MANICA, M. NEDA, AND L. REBHOLZ, *Numerical analysis and computational comparisons of the NS-alpha and NS-omega regularizations*, Computer Methods in Applied Mechanics and Engineering, 199 (2010), 916–931.

[LN06a] W. LAYTON AND M. NEDA, *Truncation of scales by time relaxation, Journal of Mathematical Analysis and Applications*, 325(2) (2007), 788–807.

[LN06b] W. LAYTON AND M. NEDA, *The energy cascade for homogeneous, isotropic turbulence generated by approximate deconvolution models*, JMAA, 333(1) (2007), 416–429.

[LPR10] W. LAYTON, C. D PRUETT AND L. G. REBHOLZ, *Temporally regularized direct numerical simulation*, Applied Mathematics and Computation, 216 (2010), 3728–3738.

[LRS10] W. LAYTON, L. REBHOLZ AND M. SUSSMAN, *Energy and helicity dissipation rates of the NS-alpha and NS-alpha-deconvolution models*, IMA Journal of Applied Mathematics 75(1) (2010), 56–74

[LS07] W. LAYTON AND I. STANCULESCU, *K-41 optimized approximate deconvolution models*, International Journal of Computing Science and Mathematics, 1 (2007), 396 - 411.

[LS09] W. LAYTON AND I. STANCULESCU, *Chebychev optimized approximate deconvolution models of turbulence*, Applied Mathematics and Computation, 208 (2009), 106–118.

[LST08] W. LAYTON, I. STANCULESCU, AND C. TRENCHEA, *Theory of the NS-$\overline{\omega}$ model*, Technical Report, University of Pittsburgh, (2008).

[LT10] W. LAYTON, AND C. TRENCHEA, *The Das-Moser commutator closure for filtering through a boundary is well posed*, Mathematical and Computer Modelling, 53, (5–6) (2011), 566–573.

[LLe06] J. LEDERER AND R. LEWANDOWSKI, *On the RANS 3D model with unbounded eddy viscosities.* Ann. IHP ann. non lin, 24(3) (2007), 413–441.

[L34a] J. LERAY, *Essay sur les mouvements plans d'une liquide visqueux que limitent des parois*, J. math. pur. appl., Paris Ser. IX, 13 (1934), 331–418.

[L34b] J. LERAY, *Sur les mouvements d'une liquide visqueux emplissant l'espace*, Acta Math., 63(1934), 193–248.

[L54] J. LERAY, *The physical facts and the differential equations*, American Math. Monthly 61 (1954), 5–7.

[Le06] R. LEWANDOWSKI, *Vorticities in a LES model for 3D periodic turbulent flows,* Journ. Math. Fluid. Mech, 8 (2006), 398–422.

[LP09] R. LEWANDOWSKI AND Y. PREAUX, *Attractors for a deconvolution model of turbulence*, Applied Mathematics Letters, 22 (2009), 642–645.

[Li01] A. LIAKOS, *Discretization of the Navier–Stokes equations with slip boundary condition*, Numerical Methods for Partial Differential Equations, 17 (2001), 26 - 42.

[Lin94] T. LINDEBERG, *Scale-space theory in computer vision*, Kluwer, Dordrecht, (1994).

[L73] J.-L. LIONS, *Perturbations singulieres dans les problemes aux limites et en controle optimal*, Springer LNM vol 323, 1973.

[LO02] G. LUBE AND M. OLSHANSKII, *Stable finite element calculations of incompressible flows using the rotation form of convection*, IMA J. Num. Anal., 22 (2002), 437–461.

[MM06] C. C. MANICA AND S. KAYA-MERDAN, *Convergence Analysis of the Finite Element Method for a Fundamental Model in Turbulence*, University of Pittsburgh Technical Report, http://www.math.pitt.edu/techreports.html, (2006).

[MNOR11] C.C. MANICA, M. NEDA, M.A. OLSHANSKII, L. REBHOLZ AND N. WILSON, *On an efficient finite element method for NS-$\overline{\omega}$ with strong mass conservation*, Computational Methods in Applied Mathematics, to appear, (2011).

[MS09] C. C. MANICA AND I. STANCULESCU, *Leray-Tikhonov regularization models of fluid motion*, University of Pittsburgh Technical Report, http://www.math. pitt.edu/techreports.html, (2009).

[Max79] J. C. MAXWELL, *On the condition to be satisfied by a gas at the surface of a solid body*, Scientific Papers 2 (1879) 704.

[MR10] W. MILES AND L. REBHOLZ, *An enhanced physics based scheme for the NS-alpha turbulence model*, Numerical Methods for Partial Differential Equations, 26(6), (2010), 1530–1555.

[MT92] H. MOFFATT AND A. TSONIBER, *Helicity in laminar and turbulent flow*, Annual Review of Fluid Mechanics 24 (1992), 281–312.

[MM87] R. MOSER AND P. MOIN, *The effects of curvature in wall bounded flows*, J. Fluid Mech. 175 (1987), 479–510.

[Mus96] A. MUSCHINSKI, *A similarity theory of locally homogeneous and isotropic turbulence generated by a Smagorinsky-type LES*, J.F.M., 325 (1996), 239–260.

[N10] M. NEDA, *Discontinuous time relaxation for the time dependent Navier–Stokes equations*, Advances in Numerical Analysis, 2010 (ID 419021) (2010), 1–21.

[O99] M.A. OLSHANSKII, *Iterative solver for Oseen problem and numerical solution of incompressible Navier–Stokes equations*, Num. Linear Algebra Appl., 6 (1999), 353–378.

[O02] M.A. OLSHANSKII, *A low order Galerkin finite element method for the Navier–Stokes equations of steady incompressible flow: A stabilization issue and iterative methods*, Comp. Meth. Appl. Mech. Eng., 191 (2002), 5515–5536.

[OR02] M. OLSHANSKII AND A. REUSKEN, *Navier–Stokes equations in rotation form: a robust multigrid solver for the velocity problem*, SIAM J. Sci. Comp., 23 (2002), 1682–1706.

[OR04] M.A. OLSHANSKII AND A. REUSKEN, *Grad-Div stabilization for the Stokes equations*, Math. Comp., 73 (2004), 1699–1718.

[OR11] M.A. OLSHANSKII AND L. REBHOLZ, *Application of barycenter refined meshes in linear elasticity and incompressible fluid dynamics*, submitted, (2011).

[P92] C. PARES, *Existence, uniqueness and regularity of solutions of the equations of a turbulence model for incompressible fluids*, Appl. Anal. 43(1992), 245–296.

[P94] C. PARES, *Approximation de la solution des equationes d'un modele de turbulence par une methode de Lagrange Galerkin*, Rev. Mat. Apl. 15(1994), 63–124.

[P08] U PIOMELLI, *Wall-layer models for Large-Eddy Simulation*, Progress in Aerospace Science, 44(2008) 437–446.

[PB02] U PIOMELLI AND E BALARAS, *Wall-layer models for Large-Eddy Simulation*, Annual Review of Fluid Mechanics 34(2002) 349–374.

[Po00] S. POPE, *Turbulent Flows*, Cambridge Univ. Press, (2000).

[Pr06] C. PRUETT, *Temporal large-eddy simulation: Theory and practice*, in: Special Issue of Large-Eddy Simulation of Complex Flows, eds. N.A. Adams and R.D. Moser, Theoretical and Computational Fluid Dynamics, 22 (3–4) (2008), 275–304.

[PGGT03] C. D. PRUETT, T. B. GATSKI, C. E. GROSCH, AND W. D. THACKER, *The temporally filtered Navier–Stokes equations: properties of the residual stress*, Phys. Fluids, 15 (2003), 2127–2140.

[PTGG06] C. D. PRUETT, B. C. THOMAS, C. E. GROSCH, AND T. B. GATSKI, *A temporal approximate deconvolution model for large-eddy simulation*, Phys. Fluids, 18 (2006), 1–4.

[Reb07] L. REBHOLZ, *Conservation laws of turbulence models*, Journal of Mathematical Analysis and Applications, 326(1) (2007), 33–45.

[RS10] L. REBHOLZ AND M. SUSSMAN, *On the high accuracy NS-alpha-deconvolution model of turbulent fluid flow*, Mathematical Models and Methods in Applied Sciences, 20(4) (2010), 611–633.

[R22] L. F. RICHARDSON, *Weather prediction by numerical process*, Cambridge University press, Cambridge, (1922).

[R89] P. ROSENAU, *Extending hydrodynamics via the regularization of the Chapman-Enskog expansion*, Phys. Rev.A, 40 (1989), 7193–7196.

[R90] N. ROTT, *Note on the History of the Reynolds Number*, Annual Review of Fluid Mechanics, 22 (1990), 1–11.

[S01] P. SAGAUT, *Large eddy simulation for Incompressible flows*, Springer, Berlin, (2001).

[ST92] S. SCHOCHET AND E. TADMOR, *The regularized Chapman-Enskog expansion for scalar conservation laws*, Arch. Rat. Mech. Anal. 119 (1992), 95–113.

[S84] K. R. SREENIVASAN, *On the scaling of the turbulent energy dissipation rate*, Phys. Fluids, 27(5) (1984) 1048–1051.

[S98] K. R. SREENIVASAN, *An update on the energy dissipation rate in isotropic turbulence*, Phys. Fluids, 10(2) (1998) 528–529.

[S08] I. STANCULESCU, *Existence theory of abstract approximate deconvolution models of turbulence*, Ann. Univ. Ferrara, 54 (2008), 145–168.

[SA99] S. STOLZ AND N. A. ADAMS, *On the approximate deconvolution procedure for LES*, Phys. Fluids, 11 (1999), 1699–1701.

[SAK01a] S. STOLZ, N. A. ADAMS AND L. KLEISER, *The approximate deconvolution model for LES of compressible flows and its application to shock-turbulent-boundary-layer interaction*, Phys. Fluids 13 (2001), 2985–3001.

[SAK01b] S. STOLZ, N. A. ADAMS AND L. KLEISER, *An approximate deconvolution model for large eddy simulation with application to wall-bounded flows*, Phys. Fluids, 13 (2001), 997–1015.

[SAK02] S. STOLZ, N. A. ADAMS AND L. KLEISER, *The approximate deconvolution model for compressible flows: isotropic turbulence and shock-boundary-layer interaction*, in: Advances in LES of complex flows (editors: R. Friedrich and W. Rodi) Kluwer, Dordrecht, (2002).

[SSK05] S. STOLZ, P. SCHLATTER, AND L. KLEISER, *High-pass filtered eddy-viscosity models for LES of transitional and turbulent flow*, Phys. Fluids, 17 (2005)065103.

[T93] L. TARTAR, *Remarks on some interpolation spaces*, in: *BVPs for PDEs and applications*, Masson, Paris, (1993), 229–252.

[TS06] J.A. TEMPLETON AND M. SHOEYBI, *Towards wall-normal filtering for large eddy simulation*, Multiscale Modeling and Simulation 5 (2006), 420–444.

[VLM98] O. VASILYEV, T. LUND AND P. MOIN, *A general class of commutative filters for LES in complex geometries*, Journal of Computational Physics, 146 (1998), 105–123.

[VG02] M. VISBAL AND D. GAITONDE, *On the use of higher order finite difference schemes on curvilinear and deforming meshes*, JCP 181 (2002) 155–185.

[VTC05] M. I. VISHIK, E. S. TITI AND V. V. CHEPYZHOV, *Trajectory attractor approximations of the 3d Navier–Stokes system by the Leray-alpha model*, Russian Math Dokladi, 71 (2005), 91–95.

[V03] A.W. VREMAN, *The filtering analog of the variational multiscale method in large-eddy simulation*, Phys Fluids 15(2003) 61–64.

[V04] A.W. VREMAN, *An eddy-viscosity subgrid-scale model for turbulent shear flow: algebraic theory and applications,* Phys. Fluids 16 (2004), 3670–3681.

[Z91] T.A. ZANG, *On the rotation and skew-symmetric forms for incompressible flow simulations*, Appl. Num. Math., 7 (1991) 27–40.

Index

W.J. Layton and L.G. Rebholz, *Approximate Deconvolution Models
of Turbulence*, Lecture Notes in Mathematics 2042,
DOI 10.1007/978-3-642-24409-4, © Springer-Verlag Berlin Heidelberg 2012

LECTURE NOTES IN MATHEMATICS

 Springer

Edited by J.-M. Morel, B. Teissier; P.K. Maini

Editorial Policy (for the publication of monographs)

1. Lecture Notes aim to report new developments in all areas of mathematics and their
   applications - quickly, informally and at a high level. Mathematical texts analysing new
   developments in modelling and numerical simulation are welcome.
   Monograph manuscripts should be reasonably self-contained and rounded off. Thus they
   may, and often will, present not only results of the author but also related work by other
   people. They may be based on specialised lecture courses. Furthermore, the manuscripts
   should provide sufficient motivation, examples and applications. This clearly distinguishes
   Lecture Notes from journal articles or technical reports which normally are very concise.
   Articles intended for a journal but too long to be accepted by most journals, usually do not
   have this "lecture notes" character. For similar reasons it is unusual for doctoral theses to
   be accepted for the Lecture Notes series, though habilitation theses may be appropriate.

2. Manuscripts should be submitted either online at www.editorialmanager.com/lnm to
   Springer's mathematics editorial in Heidelberg, or to one of the series editors. In general,
   manuscripts will be sent out to 2 external referees for evaluation. If a decision cannot yet
   be reached on the basis of the first 2 reports, further referees may be contacted: The author
   will be informed of this. A final decision to publish can be made only on the basis of the
   complete manuscript, however a refereeing process leading to a preliminary decision can
   be based on a pre-final or incomplete manuscript. The strict minimum amount of material
   that will be considered should include a detailed outline describing the planned contents
   of each chapter, a bibliography and several sample chapters.
   Authors should be aware that incomplete or insufficiently close to final manuscripts almost
   always result in longer refereeing times and nevertheless unclear referees' recommenda-
   tions, making further refereeing of a final draft necessary.
   Authors should also be aware that parallel submission of their manuscript to another
   publisher while under consideration for LNM will in general lead to immediate rejection.

3. Manuscripts should in general be submitted in English. Final manuscripts should contain
   at least 100 pages of mathematical text and should always include

   – a table of contents;
   – an informative introduction, with adequate motivation and perhaps some historical
     remarks: it should be accessible to a reader not intimately familiar with the topic
     treated;
   – a subject index: as a rule this is genuinely helpful for the reader.

   For evaluation purposes, manuscripts may be submitted in print or electronic form (print
   form is still preferred by most referees), in the latter case preferably as pdf- or zipped
   psfiles. Lecture Notes volumes are, as a rule, printed digitally from the authors' files.
   To ensure best results, authors are asked to use the LaTeX2e style files available from
   Springer's web-server at:

   ftp://ftp.springer.de/pub/tex/latex/svmonot1/ (for monographs) and
   ftp://ftp.springer.de/pub/tex/latex/svmultt1/ (for summer schools/tutorials).

Additional technical instructions, if necessary, are available on request from lnm@springer. com.

4. Careful preparation of the manuscripts will help keep production time short besides ensuring satisfactory appearance of the finished book in print and online. After acceptance of the manuscript authors will be asked to prepare the final LaTeX source files and also the corresponding dvi-, pdf- or zipped ps-file. The LaTeX source files are essential for producing the full-text online version of the book (see http://www.springerlink. com/openurl.asp?genre=journal&issn=0075-8434 for the existing online volumes of LNM). The actual production of a Lecture Notes volume takes approximately 12 weeks.

5. Authors receive a total of 50 free copies of their volume, but no royalties. They are entitled to a discount of 33.3 % on the price of Springer books purchased for their personal use, if ordering directly from Springer.

6. Commitment to publish is made by letter of intent rather than by signing a formal contract. Springer-Verlag secures the copyright for each volume. Authors are free to reuse material contained in their LNM volumes in later publications: a brief written (or e-mail) request for formal permission is sufficient.

Addresses:
Professor J.-M. Morel, CMLA,
École Normale Supérieure de Cachan,
61 Avenue du Président Wilson, 94235 Cachan Cedex, France
E-mail: morel@cmla.ens-cachan.fr

Professor B. Teissier, Institut Mathématique de Jussieu,
UMR 7586 du CNRS, Équipe "Géométrie et Dynamique",
175 rue du Chevaleret
75013 Paris, France
E-mail: teissier@math.jussieu.fr

For the "Mathematical Biosciences Subseries" of LNM:

Professor P. K. Maini, Center for Mathematical Biology,
Mathematical Institute, 24-29 St Giles,
Oxford OX1 3LP, UK
E-mail : maini@maths.ox.ac.uk

Springer, Mathematics Editorial, Tiergartenstr. 17,
69121 Heidelberg, Germany,
Tel.: +49 (6221) 4876-8259

Fax: +49 (6221) 4876-8259
E-mail: lnm@springer.com